T0205615

Springer Tracts in Advanced Robotics 122

More information about this series at http://www.springer.com/series/5208

Emilio Garcia-Fidalgo · Alberto Ortiz

Methods for Appearance-based Loop Closure Detection

Applications to Topological Mapping and Image Mosaicking

 Springer

Emilio Garcia-Fidalgo
Department of Mathematics
and Computer Science
University of the Balearic Islands
Palma de Mallorca
Spain

Alberto Ortiz
Department of Mathematics
and Computer Science
University of the Balearic Islands
Palma de Mallorca
Spain

ISSN 1610-7438   ISSN 1610-742X   (electronic)
Springer Tracts in Advanced Robotics
ISBN 978-3-030-09373-0   ISBN 978-3-319-75993-7   (eBook)
https://doi.org/10.1007/978-3-319-75993-7

Printed on acid-free paper

This Springer imprint is published by Springer Nature
The registered company is Springer International Publishing AG
The registered company address is: Gewerbestrasse 11, 6330 Cham, Switzerland

To my father Valentín. I hope you read these lines, wherever you are.

Foreword

Robotics is undergoing a major transformation in scope and dimension. From a largely dominant industrial focus, robotics is rapidly expanding into human environments and vigorously engaged in its new challenges. Interacting with, assisting, serving, and exploring with humans, the emerging robots will increasingly touch people and their lives.

Beyond its impact on physical robots, the body of knowledge robotics has produced is revealing a much wider range of applications reaching across diverse research areas and scientific disciplines, such as biomechanics, haptics, neurosciences, virtual simulation, animation, surgery, and sensor networks among others. In return, the challenges of the new emerging areas are proving an abundant source of stimulation and insights for the field of robotics. It is indeed at the intersection of disciplines that the most striking advances happen.

The *Springer Tracts in Advanced Robotics (STAR)* is devoted to bringing to the research community the latest advances in the robotics field on the basis of their significance and quality. Through a wide and timely dissemination of critical research developments in robotics, our objective with this series is to promote more exchanges and collaborations among the researchers in the community and contribute to further advancements in this rapidly growing field.

The monograph by E. Garcia-Fidalgo and A. Ortiz is based on the first author's Ph.D. thesis work. It deals with the mapping and localization problem for autonomous mobile robot navigation in cluttered environments, adopting methods for appearance-based loop closure detection. Novel topological mapping algorithms are proposed and experimentally tested on state-of-the-art datasets. A fast multi-threaded image mosaicking algorithm is also proposed as a particular case of a topological map, leading to successful results for motion estimation.

STAR is proud to welcome yet another volume in the series dedicated to the popular area of SLAM!

Naples, Italy
December 2017

Bruno Siciliano
STAR Editor

Preface

Mapping and localization are two essential processes in autonomous mobile robotics since they are the basis of other higher level and more complex tasks, such as obstacle avoidance or path planning. Mapping is the process through which a robot builds its own representation of the environment when a map of the environment is not available. In this context, there exist mainly two types of maps: metric and topological. While metric maps represent the world as accurate as possible with regard to a global coordinate system, topological maps represent the environment in an abstract manner by means of a graph, which implies several benefits in front of the classic metric approaches.

Regardless of the type of map to be built, any sensor used to perceive the environment presents an unavoidable noise that should be taken into account in order to not compromise the accuracy of the resulting map and the localization process. For this reason, an additional input is required to correct the inefficiencies produced in the map by this noise. Due to this reason, mapping algorithms usually rely on loop closure detection techniques, which entails the correct identification of previously seen places to reduce the uncertainty of the resulting maps.

In the last decades, there has been a significant increase in the number of visual solutions to solve the loop closure detection problem, specially motivated by the low cost of cameras, the richness of the sensor data provided, and the existence of more powerful computers. The problem of identifying if the current received image has been taken from an already visited place is usually known as *appearance-based loop closure detection*. In general, given that this problem deals with sets of images, any technique proposed to solve it should be devised to efficiently manage a huge volume of visual information, which normally grows along with the map.

This book deals with the problem of generating topological maps of the environment using efficient appearance-based loop closure detection techniques. Since the quality of a visual loop closure detection algorithm is related to the image description method and its ability to index previously seen images, several methods for loop closure detection adopting different approaches are developed and assessed. Then, these methods are used as basic components in three novel topological mapping algorithms. The results obtained indicate that the solutions proposed attain a better performance than several state-of-the-art approaches. To conclude, given that loop closure detection is also a key component in other research areas, a multi-threaded image mosaicking algorithm is proposed, which can be seen as a particular case of a topological map. This approach makes use of one of the loop closure detection techniques previously introduced in the book in order to find overlapping pairs between images and finally obtain seamless mosaics of different environments in a reasonable amount of time.

The work reported in this book was supported by the European Social Fund through Ph.D. scholarship FPI11-43123621R (Conselleria d'Educació, Cultura i Universitats, Govern de les Illes Balears), by FP7 projects MINOAS (GA 233715) and INCASS (GA 605200), and by H2020 project ROBINS (GA 779776).

Palma de Mallorca, Spain   Emilio Garcia-Fidalgo
December 2017   Alberto Ortiz

Contents

Acronyms

ABLC	Appearance-based loop closure
ABLE	Able for Binary-appearance loop closure evaluation
ABTM	Appearance-based topological mapping
ANNS	Approximate nearest neighbor search
AUC	Area under the curve
AUV	Autonomous underwater vehicle
BIMOS	Binary descriptor-based image mosaicking
BINMap	Binary mapping
BoW	Bag-of-words
BRIEF	Binary robust independent elementary features
BRISK	Binary robust invariant scalable keypoints
CBIR	Content-based image retrieval
CENTRIST	Census transform histogram
DFT	Discrete Fourier transform
DOG	Difference of Gaussians
FAB-MAP	Fast appearance-based mapping
FACT	Fast adaptive color tags
FEATMap	Feature-based mapping
FN	False negative
FP	False positive
FREAK	Fast retina keypoint
GPS	Global positioning system
GPU	Graphics processor unit
HCT	Hull census transform
HMM	Hidden Markov model
HOUP	Histogram of oriented uniform patterns
HTMap	Hierarchical topological mapping
IMU	Inertial measurement unit
KLT Tracker	Kanade–Lucas–Tomasi tracker
LDB	Local difference binary

LTM	Long-term memory
LSH	Locality-sensitive hashing
LVQ	Learning vector quantization
MAV	Micro-aerial vehicle
MRF	Markov random field
NNS	Nearest neighbor search
OACH	Orientation adjacency coherence histograms
OBIndex	Online binary image index
OpenCV	Open-source computer vision
ORB	Oriented FAST and rotated BRIEF
PCA	Principal component analysis
PHOG	Pyramid histogram of oriented gradients
PIRF	Position-invariant robust feature
PR Curve	Precision–recall curve
PTM	Probabilistic topological maps
RANSAC	Random sample consensus
ROC	Receiver operating characteristic
ROS	Robot operating system
SAT	Separating axis theorem
SeqSLAM	Sequence SLAM
SIFT	Scale-invariant feature transform
SLAM	Simultaneous localization and mapping
STM	Short-term memory
SURF	Speeded-up robust features
SVM	Support vector machines
TF-IDF	Term frequency–inverse document frequency
TN	True negative
TP	True positive
U-SURF	Upright-SURF
VPC	Visual place categorization
WGII	Weighted grid integral invariant
WGOH	Weighted gradient orientation histograms

Symbols

t	Time stamp t
I_t	Image at time t
F_t	Set of local feature descriptors extracted from image I_t
f_j^t	Local feature descriptor j belonging to set F_t
G_t	Global descriptor computed from image I_t
$d_f(f_p^i, f_q^j)$	Generic distance between two local feature descriptors
$d_g(G_i, G_j)$	Generic distance between two global descriptors
M_t	Topological map at time t
γ	Graph encoding a topological map
ω	Generic set of nodes
β	Generic index of images
κ_i	Generic keyframe
τ_i	Generic threshold
c	Loop closure candidate index
p	Number of recent images discarded as loop closure candidates
z_t	Single observation at time t
O_t	Set of observations at time t
L_i^t	Event that image I_t closes a loop with image I_t
η	Normalizing factor
s	Generic score
P_j^i	Path between nodes i and j
$E(P_j^i)$	Erasability of path P_j^i
ρ	Nearest neighbor distance ratio
ℓ_i	Generic location
ϕ_i	Representative global descriptor of the location ℓ_i
p_i	Point in an image
^iH_j	Absolute homography between frames i and j

$^iH_j^*$	Relative homography between frames i and j
^iO_j	Overlap between images I_i and I_j
ε	Error function
$h(\varepsilon)$	Huber robust error function
$R(^iH_j)$	Regularization term associated with homography ^iH_j

List of Figures

List of Tables

List of Algorithms

Chapter 1
Introduction

Abstract This chapter serves as an introduction to this book. In the first section, we briefly consider some basic concepts from the robotics field, such as navigation, mapping, localization and SLAM. We also comment on the concept of appearance-based loop closure detection and discuss its importance for mapping and localization tasks. The main contributions are outlined next, once the scope of the book has already been stated.

1.1 Motivation

1.1.1 Basic Concepts

Robotics is a research field that has gained much popularity in recent years due to the growing interest of big companies. This increasing popularity has also promoted the creation of new corporations dedicated to particular markets in robotics. As a consequence of these facts, robotics has evolved from a purely academic field to be present in the people's daily lives. Robotic applications covers an extensive range of possibilities in different areas, such as, for instance, surgery, space exploration, surveillance, security, personal assistance, inspection of structures or rehabilitation. Among other definitions, robotics could be stated as the branch of engineering that involves the design, manufacture, control and programming of *robots*. Robotics takes concepts from different subjects such as mechanical engineering, electrical engineering, electronic engineering, mathematics, physics and computer science, and, therefore, it can be seen as a combination of several areas of knowledge. Figure 1.1 illustrates some examples of different robots developed recently.

The term *robot* was first coined by the Czech writer Karel Capek in his science fiction play Rossum's Universal Robots (R.U.R). It comes from the Czech word *robota*, which means slave or forced labour. It was popularized by Isaac Asimov in 1950 in his novel *I Robot*, where, by the way, the Three Laws of Robotics were introduced. Nowadays, there is no clear consensus on what the term means, since it is used to designate agents that perform different tasks in many kinds of scenarios.

© Springer International Publishing AG 2018

E. Garcia-Fidalgo and A. Ortiz, *Methods for Appearance-based Loop Closure Detection*, Springer Tracts in Advanced Robotics 122, https://doi.org/10.1007/978-3-319-75993-7_1

Fig. 1.1 Examples of modern robots. **a** Google's self-driving car (photo by Michael Shick / CC BY-SA 4.0). **b** Waymo's self-driving car (photo by Grendelkhan / CC BY-SA 4.0). **c** DJI Phantom 3 (photo by Jacek Halicki / CC BY-SA 4.0). **d** Parrot's AR.Drone. **e** iRobot's Roomba 780 vacuum cleaner (photo By Tibor Antalóczy / CC BY-SA 3.0). **f** Asctec's Falcon 8 drone (photo by Eugehal / CC BY-SA 4.0). All images were obtained via Wikimedia Commons

In attempting to give a modern definition, we could say that robots are machines which are able to execute one or more tasks repeatedly while interact with their environment.

In order to perform more complex tasks, robots usually need to be moved, which leads us to a special kind of robots that are *mobile*. Therefore, a *mobile robot* is an automatic machine that has the ability to move around in their environment and it is not fixed to a physical location. The type of movement of a mobile robot is determined according to their locomotion mechanisms. Although a lot of actuators have been proposed by roboticists along the years, the election of the locomotion mechanisms to be incorporated into a mobile vehicle is normally governed by the domain of the application. According to Dudek and Jenkin [1], mobile robots can be broadly classified into four main categories conforming to this application domain:

- *Terrestrial robots*, which walk on the ground and are designed to take advantage of a solid contact with a surface. Initially, the most common terrestrial vehicles were wheeled, but recently, there exists a growing interest in humanoid robots, which are mainly based on bipedal locomotion.
- *Aquatic robots*, which operate in water, either at the surface or underwater. This kind of robots is mainly equipped with water jets or propellers as locomotion mechanisms. The importance of this group of robots lies in that water, specially in scenarios like oceans, is a medium usually hard to reach for humans.
- *Aerial robots*, which can fly. They share many issues and locomotion mechanisms with aquatic robots. This kind of robot has become an important research field in the last few years due to the intense development of Micro Aerial Vehicles (MAV).

- *Space robots*, which are designed to operate in the microgravity of outer space. The locomotion strategies used for these robots vary, given that some of them are mainly devised for flying and others for ground exploration.

A mobile robot can be managed by a human operator. However, there are some tasks where robots are required to be *autonomous*. By definition, an *autonomous robot* is an agent that can perform behaviours or tasks with a high degree of autonomy and/or without human intervention. This kind of robots is particularly interesting in fields such as space exploration or rescuing tasks in disasters, which are hard to reach for humans.

1.1.2 Mobile Robot Navigation

When converting a mobile robot in an autonomous platform, several problems arise. In addition to the locomotion mechanisms required for the task at hand, the robot needs to perceive information from its environment and process it in a intelligent way in order to plan routes, reach places and avoid dangerous situations. In this context, these tasks lead us to another important concept in mobile robotics: *navigation*.

Mobile robot navigation can be roughly described as the process of determining a suitable and safe path between a starting and a goal point for a robot travelling between them [2]. Navigation can be seen as a combination of, among others, three fundamental tasks:

- *Localization*, which denotes the ability of the robot to know its position and orientation with regard to its environment.
- *Path planning*, which deals with finding the most optimal path between two points of the environment. Note that this task usually includes the ability of avoiding obstacles.
- *Mapping*, which is in charge of creating *maps*, abstract representations of the robot's environment, using the sensors the robot is equipped with.

Given the nature of each of these tasks, we can argue that maps are essential components in most robotic navigation systems, since the other tasks depend significantly on them. For instance, path planning, to be correctly carried out, require a map of the environment and the location of the robot within the map. It is also true that some authors have proposed reactive navigation techniques encompassed in a category called *mapless* navigation systems [2], where there is no global representation of the environment and the world is perceived as the system navigates through it. Nonetheless, most part of the solutions proposed during the last years belong to the group of *map-based* navigation systems, where a map of the environment is needed to navigate.

1.1.3   Mapping, Localization and SLAM

Maps could be provided to a robot a priori, but sometimes this is not possible, and then the robot is required to build its own representation of the possibly unknown or partially unknown environment. This process, as mentioned previously, is called *mapping* and is one of the main topics of this book. As far as robotic mapping is concerned, two main paradigms are generally accepted: metric and topological mapping. While metric maps describes the position of the robot in the world along with the detected objects according to a global coordinate system, topological maps represent the environment in an abstract manner by means of a graph. Advantages and disadvantages of each paradigm will be discussed in Chap. 2. In this book, we are primarily concerned with finding methods for generating topological maps of the environment.

Despite mapping and localization can be performed as independent tasks, they are closely related. As a result of the mapping process, a representative map of the environment is generated while the localization process computes the pose of the robot within this map according to the sensor data perceived from the sensors. As mentioned above, both processes can be used for navigation-related tasks and are of special interest for autonomous vehicles, which need to be able to operate without any human intervention. The pose of the structures and the obstacles of the environment needs to be known to build a map. Whereas, during localization, the pose of the agent against a reference map is computed. In this case the map of the working scenario must be available before starting the navigation, which limits the autonomy of the vehicle. To solve this egg-and-hen problem, several approaches have been proposed where both tasks take place at the same time, creating an incremental map of an unknown environment while localizing the robot within this map. These techniques are generically known as Simultaneous Localization and Mapping (SLAM) [3].

1.1.4   Loop Closure Detection

One of the most important aspects of both mapping and localization tasks and, by association, SLAM, is to correctly perceive the information of the environment. Different kinds of sensors, such as sonars, radars, laser range finders or cameras have been used during years for this purpose. However, none of them are exempt from noise. This unavoidable noise produces inaccuracies in localization and mapping tasks, leading to inconsistent representations of the environment if only raw sensor data is considered. Due to this reason, mapping approaches are heavily influenced by *loop closure* detection. An example of the effect of loop closure detection in mapping tasks is illustrated in Fig. 1.2.

Loop closure detection, usually also known as *place recognition*, is a key challenge to overcome which entails the correct identification of previously visited places from sensor data. This allows the robot to correct inaccuracies in the map, generating

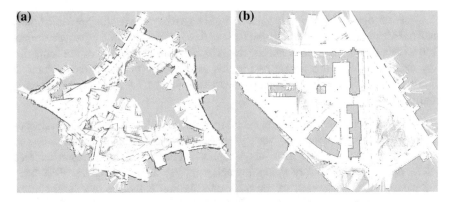

Fig. 1.2 Example of metric map generated using a laser range finder. **a** Resulting map using raw sensor data. **b** The same map improved using, among other information, the detected loop closures for correcting the trajectory. These images result from the *Freiburg Campus* dataset. This dataset was obtained from the Robotics Data Set Repository (Radish) [4]. Thanks go to Cyrill Stachniss and Giorgio Grisetti [5] for providing this data

consistent representations and reducing their uncertainty. However, as stated by Korrapati [6], loop closure detection is not an easy task and is even a more complex problem than localization due to the following reasons:

- *Scalability*: The complexity of the problem increases as the map enlarges, since more previous observations need to be compared with the current one to determine the existence of a loop.
- *Perceptual aliasing*: Different places of the environment are perceived as the same, resulting into false loop closure detections.
- *Sensor noise*: As stated above, measurements include noise, which makes more difficult the data association.
- *Changes in the environment*: Moving objects produce different perceptions of the same place.

Recently there has been a significant increase in the number of visual solutions to solve the loop closure detection problem because of the low cost of cameras and the richness of the sensor data provided. This implies to carry out the loop closure detection using *images* as a main source of information.

1.2 Contributions

Due to the importance of place recognition in mapping tasks, this book introduces several appearance-based loop closure detection techniques devised from different points of view. Then, using these techniques as a basis, the rest of the book is concerned with the development of a set of vision-based topological mapping algorithms.

However, recognizing previously seen places is also a key step in other research fields, such as image mosaicking, which can be seen as a special case of a topological map. Due to this reason, an image mosaicking algorithm based on one of the loop closure detection techniques introduced in this work is also proposed, as an example of application of loop closure in a different research area. More precisely, the main contributions of this book are:

- A complete survey of the most outstanding works in vision-based topological mapping and localization methods during the last fifteen years. This survey has allowed us to determine several open research topics that inspire the rest of contributions.
- The development of a new topological mapping algorithm called *FEATMap* (Feature-based Mapping), whose main contribution is the use of a loop closure detection module based on a set of randomized kd-trees and inverted files for indexing previously seen images. Another contribution of FEATMap is a map refinement strategy to avoid redundant nodes in the resulting maps and to refine the final estimated topology.
- A novel technique for indexing images called *OBIndex* (Online Binary image Index). This method is based on an incremental Bag-of-Binary-Words scheme, taking advantage of the benefits of binary descriptors and avoiding the common training step of Bag-of-Words approaches. This image indexing scheme is a key component for the remaining contributions of this book.
- A new topological mapping algorithm called *BINMap* (Binary Mapping). As a main innovation, BINMap utilizes OBIndex to index the previously seen images and to obtain similar loop closure candidates in an efficient way, achieving good recall rates.
- A novel vision-based approach for topological mapping called *HTMap* (Hierarchical Topological Mapping), which relies on a hierarchical loop closure detection algorithm. In HTMap, images are described using a global descriptor and a set of binary local descriptors, and similar images are grouped together to form a location. Each location is represented by means of an average global descriptor and an instance of OBIndex containing the descriptors found in the images associated to the location. Then, the loop closure detection is performed in two different steps: first, the global descriptors are used to obtain candidate locations, and, secondly, the instances of OBIndex of each node are used for obtaining similar image candidates in the retrieved nodes.
- A novel image mosaicking algorithm called *BIMOS* (Binary descriptor-based Image Mosaicking), which can generate mosaics in a reasonable amount of time under different operating conditions. In order to efficiently estimate the topology of the environment, BIMOS uses OBIndex as a place recognition system.

Some parts of this book, which describe the aforementioned contributions, can also be found in [7–14].

1.3 Book Organization

With the preceding contributions in mind, the book is divided into nine chapters as follows:

- **Chapter** 2 reviews basic concepts and background for this book.
- **Chapter** 3 extensively discusses the main contributions on vision-based topological mapping emerged during the last fifteen years, and identifies the strong and weak points of the different approaches.
- **Chapter** 4 explains a common framework used to validate the topological mapping algorithms presented in this book. Performance metrics, datasets and reference solutions taken as baseline for comparisons are discussed.
- **Chapter** 5 introduces FEATMap, a probabilistic topological mapping approach based on local image features and a map refinement strategy.
- **Chapter** 6 introduces OBIndex, an image indexing approach based on an incremental Bag-of-Binary-Words scheme. Then, we introduce BINMap, a topological mapping algorithm which employs BINMap as a base component for detecting loop closures.
- **Chapter** 7 introduces HTMap, an appearance-based approach for topological mapping based on a hierarchical decomposition of the environment.
- **Chapter** 8 describes BIMOS, a multi-threaded approach for fast image mosaicking, which uses OBIndex as image index for inferring the relationships between the images that conform the topology of the sequence.
- **Chapter** 9 concludes the book by summarizing the main contributions achieved and by highlighting the differences of the introduced approaches with other similar solutions. Some future work to extend the research described here is also suggested.

References

1. Dudek, G., Jenkin, M.: Computational Principles of Mobile Robotics. Cambridge University Press, Cambridge (2010)
2. Bonin-Font, F., Ortiz, A., Oliver, G.: Visual navigation for mobile robots: a survey. J. Intell. Robot. Syst. **53**(3), 263–296 (2008)
3. Durrant-Whyte, H., Bailey, T.: Simultaneous localisation and mapping (SLAM): part I the essential algorithms. IEEE Robot. Autom. Mag. **2**, 99–110 (2006)
4. Howard, A., Roy, N.: The Robotics Data Set Repository (Radish) (2003). http://radish.sourceforge.net/
5. Stachniss, C., Grisetti, G., Hähnel, D., Burgard, W.: Improved rao-blackwellized mapping by adaptive sampling and active loop-closure. In: Workshop on self-organization of adaptive behavior (SOAVE), pp. 1–15 (2004)
6. Korrapati, H.: Loop closure for topological mapping and navigation with omnidirectional images. Ph.D. thesis, Université Blaise Pascal-Clermont-Ferrand II (2013)
7. Garcia-Fidalgo, E., Ortiz, A.: Hierarchical place recognition for topological mapping. IEEE Trans. Robot. **33**(5), 1061–1074 (2017)
8. Garcia-Fidalgo, E., Ortiz, A.: Vision-based topological mapping and localization methods: a survey. Rob. Auton. Syst. **64**, 1–20 (2015)

9. Garcia-Fidalgo, E., Ortiz, A.: Vision-based topological mapping and localization by means of local invariant features and map refinement. Robotica **33**(7), 1446–1470 (2015)
10. Garcia-Fidalgo, E., Ortiz, A., Bonnin-Pascual, F., Company, J.P.: Fast image mosaicing using incremental bags of binary words. In: IEEE international confernce robotics automation, pp. 1174–1180 (2016)
11. Garcia-Fidalgo, E., Ortiz, A., Bonnin-Pascual, F., Company, J.P.: A mosaicing approach for vessel visual inspection using a micro aerial vehicle. In: IEEE/RSJ international conference intelligent robots system (2015)
12. Garcia-Fidalgo, E., Ortiz, A.: On the use of binary feature descriptors for loop closure detection. In: IEEE emerging technology and factory automation, pp. 1–8 (2014)
13. Garcia-Fidalgo, E., Ortiz, A.: Probabilistic appearance-based mapping and localization using visual features. In: IBPRIA, pp. 277–285. Funchal (Portugal) (2013)
14. Garcia-Fidalgo, E., Ortiz, A.: Indexing invariant features for topological mapping and localization. In: Workshop on field robotics (euRathlon/ARCAS) (2014)

Chapter 2
Background

Abstract This chapter is intended to provide the reader with a general overview of the most important concepts and terms needed to understand the rest of the book. Main concepts are briefly introduced, making use of examples as they are needed for illustration purposes. More precisely, in the first section, we consider the concept of topological map and define it in a formal way, as well as discuss its main advantages and disadvantages in front of metric approaches. Next, we deal with appearance-based loop closure detection and the factors that more affect the performance of the underlying algorithms.

2.1 Topological Mapping

As previously stated in Chap. 1, *robot mapping* can be defined as the process of generating a representation of the environment useful for the task at hand. Despite this process seems to be an easy task for humans, it is not for a robot and therefore mapping is currently a very active research field. An appropriate map for an autonomous robot should be constructed using the same sensors that the robot employs to observe the world. These sensors usually corrupted by noise and interferences, which makes more difficult the data association between measurements and map.

There exist several ways of representing a map. The accuracy of the map will depend on the information requirements of the application. Three main paradigms for mobile robot mapping are usually accepted:

- *Metric maps*: this kind of maps represents the world as accurate as possible. They maintain a high amount of information about environment details, such as distances, measures, sizes and so on, and they are referenced according to a global coordinate system. The main drawbacks of this approach are the processing time and the storage needs, which makes its use in some real time applications more difficult.
- *Topological maps*: this approach generates an abstract representation of the world, usually as a graph with nodes and links between them. Nodes represent environment locations with similar features and links are relationships or possible actions

© Springer International Publishing AG 2018

E. Garcia-Fidalgo and A. Ortiz, *Methods for Appearance-based Loop Closure Detection*, Springer Tracts in Advanced Robotics 122, https://doi.org/10.1007/978-3-319-75993-7_2

Table 2.1 Advantages and disadvantages of metric and topological maps

Feature	Metric maps	Topological maps
CPU Needs	High	Low
Storage needs	High	Low
Path planning	Complex	Simple
Optimal routes	Yes	No
Accuracy	Yes	No

to take between the different locations. These maps are simpler and more compact than metric maps, and require much less space to be stored.

- *Hybrid maps*: this last paradigm tries to maximize the advantages and minimize the problems of each kind of map alone and combine them in a different mapping technique.

The main advantages and disadvantages of metric and topological approaches [1] are summarized in Table 2.1. As mentioned above, topological maps generally require less storage space and are more computationally efficient than metric maps, due to a simpler representation of the environment. This representation also reduces the navigation problem to finding a path between two nodes, which can be solved by any graph search algorithm. However, unlike metric maps, these paths are not always the optimal one between the nodes, due to the lack of geometric information. Besides, topological maps are not useful for tasks with accuracy needs, such as obstacle avoidance.

Given the advantages that topological maps present, in this book we are interested in developing techniques to generate this kind of maps. Formally, a graph-based map can be described as [2]:

$$G = (V, E),\qquad(2.1)$$

where V is a set of nodes and E is a set of edges. The set of n nodes is denoted as:

$$V = \{v_1, \ldots, v_n\},\qquad(2.2)$$

and the set of m edges is denoted as:

$$E = \{e_1, \ldots, e_m\},\qquad(2.3)$$

where and edge e_{ij} is expressed as:

$$e_{ij} = \{v_i, v_j\}.\qquad(2.4)$$

When the order of v_i and v_j is significant, the edge between nodes is unidirectional, resulting into a *directed* graph. Otherwise, the edges are valid for both directions, resulting into an *undirected* graph. The topological mapping solutions presented

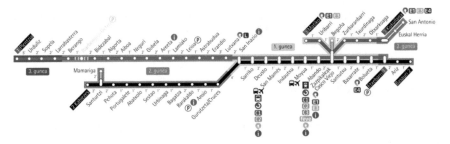

Fig. 2.1 Classical example of a topological map: Metro Bilbao network. Note that, in this kind of maps, no metric information is available, but it is enough for human orientation. Image by Laukatu/CC BY-SA 3.0, via Wikimedia Commons

in this book belong to this last type of graphs. An example of a topological map commonly used by humans is shown in Fig. 2.1.

Using vision as a sensor to generate topological maps is commonly referred to as Appearance-Based Topological Mapping (*ABTM*) [3] and is the main theme of this book. Images provide rich information and besides a camera is a cheaper sensor in comparison with other solutions, such as laser range finders.

There exist mainly two types of topological maps: dense and sparse. In a dense topological map, each new received image is added to the map as a new node. Conversely, a sparse topological map tries to group similar images together in nodes following a similarity criterion. In this book, solutions of both approaches are proposed.

2.2 Appearance-Based Loop Closure Detection

In topological mapping, loop closure detection provides information about whether the current node should be linked in the graph with a previously seen node or not. Hence, the accuracy of the final map will depend on this loop closure detection process, and wrong detections will result into inconsistent maps.

Metric information can be useful to infer or reduce the areas of the environment to search for loop closures. Given that pure topological mapping approaches do not rely on an odometry source, the loop closure detection must be performed using the sensors onboard the platform. In this regard, when a camera is the elected sensor, loop closure detection approaches are usually known as Appearance-Based Loop Closure detection (*ABLC*). The topological mapping approaches presented in this book heavily rely on ABLC. Due to this reason, several loop closure detection techniques have been developed.

The performance of an appearance-based loop closure detection algorithm is highly influenced by the description method used to describe images and the ability of the algorithm to retrieve images similar to the current one. In the following, we will perform a brief introduction of these main factors.

2.2.1 Image Description

In this section, we provide a general overview of some image description methods which are relevant for this book. These methods can be classified in two main categories: global descriptors and local image features.

2.2.1.1 Global Descriptors

Overview

Global descriptors describe the image in a holistic manner, using the full image as input to the process. These descriptors are normally very fast to compute, what simplifies the matching process between images and reduces the computational needs of mapping and localization tasks. This kind of descriptor has been used in several applications comprising scene classification, giving reasonable results in all cases.

A summary of global descriptors used in some topological mapping approaches is shown in Table 2.2. There exist other global descriptors that have not been included in the table because, to the best of our knowledge, they have not been employed in topological mapping and localization solutions, although they could be interesting for the reader. Some of them have been used for scene categorization, such as Census Transform Histogram (CENTRIST) [4], Pyramid Histogram of Oriented Gradients (PHOG) [5], Histogram of Oriented Uniform Patterns (HOUP) [6], Multi-resolution BoW [7] and for pedestrian detection, such as Histogram of Oriented Gradients (HOG) [8].

PHOG

Pyramid Histogram of Oriented Gradients (PHOG) [5] was originally developed for scene classification. It is briefly introduced here since is a key component of one of the solutions presented in this monograph. To the best our knowledge, it is the first attempt of using the PHOG descriptor for topological mapping.

PHOG represents an image by its local shape and the spatial layout of this shape. The local shape is represented by a histogram of edge orientations (HOG) [8] quantized into K bins computed from an image subregion, where the contribution of each edge is weighted according to the magnitude of the gradient. The spatial layout is represented by tiling the image into regions at multiple resolutions. More precisely, the image is split into L levels, dividing the image into a sequence of increasingly finer spatial grids by repeatedly doubling the number of divisions in each direction.

In order to compute the descriptor, a HOG vector is calculated for each grid cell at each pyramid resolution level. Then, the final PHOG descriptor is created

Table 2.2 Summary of global image descriptors

Name	References
Principal components	[9, 10]
Colour histograms	[11]
Gradient orientation Histograms	[12]
WGOH	[13]
WGII	[14]
OACH	[15]
Receptive field histograms	[16]
Gist	[17]
Omni-Gist	[18]
BRIEF-Gist	[19]
Spherical harmonics	[20]
Fingerprints	[21]
FACT	[22]
DP-FACT	[23]
Fourier signatures	[24, 25]
Colour segmented images	[26]
Scanline intensity profile	[27]
Normalized patches	[28]
2D Haar wavelet decomposition	[29, 30]
WI-SURF	[31]
WI-SIFT	[31]
DIRD	[32]
OFM	[33]
OFSC	[48]

by concatenating all the HOG descriptors and normalizing them to sum to unity. A graphical representation of the descriptor is shown in Fig. 2.2.

2.2.1.2 Local Image Features

Overview

Global descriptors, where the image is described using the entire image content, work well for capturing the general structure of the scene, but they are not able to cope with several visual problems like partial occlusions, illumination changes or camera rotations. These problems have been addressed more intensively through the development of local image features.

Fig. 2.2 Example of PHOG descriptor computation: from left to right, grids for levels 0, 1 and 2; the final descriptor consists of a weighted concatenation of the histograms of oriented gradients obtained for each grid cell

Local image features, usually known as *keypoints*, are defined as interest points in the image which are different enough from their neighbours, according to different criteria such as colour, texture and so on. During the *extraction* step, a set of distinctive local features, which capture the essence of the image, are detected. These features can be derived from the application of a neighbourhood operation or searching for specific structures within the image, such as corners, blobs or regions. Then, a *description* step is performed, where some measurements are taken from the vicinity of each local feature to form a descriptor. Initially, descriptors were formed as multi-dimensional floating-point vectors. Recently, several authors have proposed binary descriptors, where local features are defined as bit strings, reducing the storage and computational needs.

In order to identify the same local features in other images, keypoints need to be invariant to certain properties, such as camera rotations or affine transformations. According to [34], a good feature detector should have the following properties: repeatability, distinctiveness, locality, quantity, accuracy and efficiency. The most important property is repeatability, that can be achieved either by invariance, when large deformations are expected because of relevant viewpoint changes, or by robustness, in case of relatively small deformations.

Tables 2.3 and 2.4 collect relevant information about main feature detectors and descriptors. In Table 2.3, detectors are classified based on the type of the feature extracted following the guidelines of [34], where they distinguished between corner, blob and region detectors. The descriptors summarized in Table 2.4 are classified according to their type (floating-point or binary). The descriptor size, in number of components, is also shown in the table. These tables do not intend to be complete, but a summary of the most important facts about local feature detection and description. A brief overview of the approaches which are relevant for this book is given in the following. Note that some approaches comprise both feature detection and description, while others are only devised to perform one of these tasks. The interested reader is referred to [34–37] for further information about local image features.

Table 2.3 Summary of local feature detectors. Check marks between parentheses indicate that there exist versions that are invariant to scale or affine transformations

Name	References	Type of detector	Invariant		
			Rotation	Scale	Affine
Harris	[38]	Corners	✓	(✓)	(✓)
Shi and Tomasi	[39]	Corners	✓		
SUSAN	[40]	Corners	✓		
FAST	[41]	Corners	✓	(✓)	
FAST-ER	[42]	Corners	✓	(✓)	
ORB	[43]	Corners	✓	✓	
AGAST	[44]	Corners	✓	(✓)	
BRISK	[45]	Corners	✓	✓	
SIFT	[46]	Blobs	✓	✓	
SURF	[47]	Blobs	✓	✓	
CenSure	[48]	Blobs	✓	✓	
Star	[49]	Blobs	✓	✓	
SUSurE	[50]	Blobs	✓	✓	
KAZE	[51]	Blobs	✓	✓	
AKAZE	[52]	Blobs	✓	✓	
ASIFT	[53]	Blobs	✓	✓	✓
MSER	[54]	Regions	✓	✓	✓

SIFT

Scale-Invariant Feature Transform (SIFT) is an algorithm developed by Lowe [46] to detect and describe distinctive keypoints in images which was originally created for object recognition. These keypoints are invariant to image rotation and scale and robust across affine distortion, noise and changes in illumination. To obtain these keypoints, a scale space is generated convolving the original image with Gaussian kernels at different scales. A set of Difference of Gaussians (DoG) images is obtained subtracting the successive blurred images. Key locations are defined as maxima and minima of the DoGs that occur at multiple scales (see Fig. 2.3). Specifically, a DoG image D is given by:

$$D(x, y, \sigma) = L(x, y, k_i\sigma) - L(x, y, k_j\sigma), \qquad (2.5)$$

where $L(x, y, k\sigma)$ is the convolution of the original image I with a Gaussian kernel G at scale $k\sigma$:

$$L(x, y, k\sigma) = G(x, y, k\sigma) * I(x, y). \qquad (2.6)$$

Scale-space extrema detection produces too many candidates, some of which are unstable. Therefore, the next step is to perform a filtering process using the quadratic

Table 2.4 Summary of local feature descriptors

Name	References	Component type	Number of components	Invariant		
				Rotation	Scale	Affine
SIFT	[46]	Float	128	✓	✓	
SURF	[47]	Float	32, 64, 128	✓	✓	
U-SURF	[47]	Float	32, 64, 128		✓	
GLOH	[37]	Float	64, 128	✓	✓	
PCA-SIFT	[55]	Float	36	✓	✓	
M-SIFT	[56]	Float	128	✓	✓	
DAISY	[57]	Float	200	✓	✓	
LESH	[58]	Float	128	✓	✓	
ASIFT	[53]	Float	128	✓	✓	✓
KAZE	[51]	Float	64	✓	✓	
BRIEF	[59]	Bit	128, 256, 512			
ORB	[43]	Bit	256	✓	✓	
BRISK	[45]	Bit	512	✓	✓	
FREAK	[60]	Bit	512	✓	✓	
AKAZE	[52]	Bit	488	✓	✓	
D-BRIEF	[61]	Bit	32	✓	✓	
LDAHash	[62]	Bit	128	✓	✓	
BinBoost	[63]	Bit	64	✓	✓	
LDB	[64]	Bit	256, 512	✓	✓	
CBDF	[65]	Bit	256	✓	✓	

Taylor expansion of the DoG scale-space function and the eigenvalues of the second-order Hessian matrix, resulting into a reduced set of key locations. This keypoint detection is combined with a 128-dimensional descriptor, calculated on the basis of gradient orientation histograms of 4×4 subregions around the interest point.

SURF

Speeded Up Robust Features (SURF) is an image feature detection and description algorithm presented by Bay et al. [47]. It is partly inspired by the SIFT algorithm but outperforms previous solutions in terms of computational time. The SURF detector is based on the Hessian matrix. Given a point $\mathbf{x} = (x, y)$ in an image I, the Hessian matrix in x at scale σ is defined as:

$$H(\mathbf{x}, \sigma) = \begin{bmatrix} L_{xx}(\mathbf{x}, \sigma) & L_{xy}(\mathbf{x}, \sigma) \\ L_{xy}(\mathbf{x}, \sigma) & L_{yy}(\mathbf{x}, \sigma) \end{bmatrix}, \quad (2.7)$$

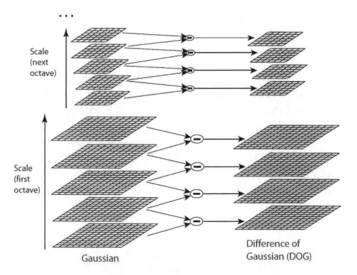

Fig. 2.3 Overview of the DoG scheme, used by SIFT for keypoint detection. For each octave, the image is convolved with Gaussian kernels to produce the set of scale spaces shown on the left. Adjacent Gaussian images are subtracted to produce the DoG images shown on the right. Image taken from [46]

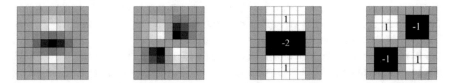

Fig. 2.4 Left to right: the Gaussian second order partial derivatives in y-direction and xy-direction, and SURF approximations using box filters. The grey regions correspond to zeros. Image taken from [47]

where $L_{xx}(\mathbf{x}, \sigma)$ is the convolution of the Gaussian second order derivative $\frac{\partial^2}{\partial x^2}G(\sigma)$ with the image I at point $\mathbf{x}$, and similarly for the rest of the terms. The determinant of this matrix is used for selecting the location and the scale. Denoting the Hessian components by D_{xx}, D_{yy} and D_{xy}, the blob response at location x in the image can be approximated by:

$$det(H_{approx}) = D_{xx}D_{yy} + (0.6D_{xy})^2. \tag{2.8}$$

These responses are stored in a blob map, and local maxima are detected and refined using a quadratic interpolation. The Hessian is roughly approximated using a set of box-type filters, as shown in Fig. 2.4. These approximations can be evaluated very fast, and independently of the image size, using integral images:

$$II(x, y) = \sum_{\substack{x' \leq x \\ y' \leq y}} I(x', y'), \qquad (2.9)$$

where $I(x, y)$ is the input image. The 9×9 box filters illustrated in Fig. 2.4 are approximations for a Gaussian with $\sigma = 1.2$ and represent the finest scale. The SURF descriptors show how the pixel intensities are distributed within the neighbourhood of each feature at different scales, resulting into a 64-dimensional vector, or a 128-dimensional vector when using the extended version. The authors also provide a faster version of the algorithm called Upright-SURF (U-SURF) where the orientation of the point is not computed and can be used in applications without rotation invariance requirements.

FAST

Features from Accelerated Segment Test (FAST) is a corner detector proposed by Rosten and Drummond [41]. It is based on the SUSAN [40] detector. FAST compares the intensity in a circle of 16 pixels around the candidate point, as shown in Fig. 2.5. Initially pixels 1 and 2 are compared with a threshold, then 3 and 4 as well as the remaining ones at the end. The pixels are classified, according to its intensity, into dark, similar and brighter groups. An image point is a corner if a minimum number of pixels can be found on the circle of fixed radius around the point such that these pixels are all brighter or darker than the central point. The feature descriptor consists of a vector containing the intensities of the 16 pixels surrounding the point, but normally this corner detector is combined with other feature descriptors. FAST has been reported as 30 times faster than a DoG detector, such as SIFT. However, it is not invariant to scale changes and it depends on a predefined threshold.

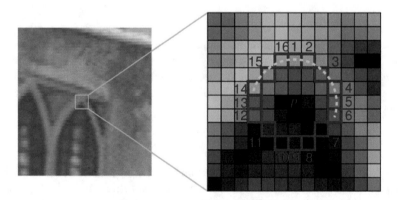

Fig. 2.5 FAST corner detection. The highlighted pixels are used by FAST for corner detection. The point p is the pixel candidate to be a corner. Image taken from [41]

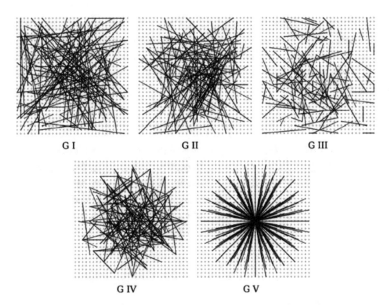

Fig. 2.6 Different spatial arrangements for computing BRIEF: G I: $(x, y) \approx U(-\frac{S}{2}, \frac{S}{2})$; G II: $(x, y) \approx N(0, \frac{1}{25}S^2)$; G III: $x \approx N(0, \frac{1}{25}S^2)$, $y \approx N(x, \frac{1}{100}S^2)$; G IV: (x, y) are randomly sampled from discrete locations of a coarse polar grid introducing a spatial quantization; G V: $\forall i : x = (0, 0)^T$ and y takes all possible values on a coarse polar grid containing the number of desired tests points; the patch size is $S \times S$ pixels and the origin of its coordinate system is located at the centre. Image taken from [59]

BRIEF

Binary Robust Independent Elementary Features (BRIEF) is a simple binary descriptor created by Calonder et al. [59]. The main goal of BRIEF is to speed up the matching process. The descriptor is a binary string, where each bit represents a simple comparison between two pixels inside a patch in the image. A bit is set to 1 if the first point has a higher intensity than the second one. In the original work, the authors suggest several point spatial arrangements over a keypoint centred patch, which are shown in Fig. 2.6. According to the authors, empirically, better results are obtained with pixel pairs randomly drawn from uniform or Gaussian distributions of point coordinates. The Hamming distance is used for matching, taking the advantage of the XOR and bit-counting CPU instructions, included in modern computers.

ORB

Oriented FAST and Rotated BRIEF (ORB) is an image feature detector and descriptor which combines the two corresponding techniques into a newer version given their good general performance as independent solutions. Keypoints are detected using a multi-scale version of FAST and filtered according to the Harris corner measure.

For each corner, an orientation is computed using the intensity centroid of the patch. Then, a steered BRIEF descriptor is computed according to the orientation of the corner. The authors use Locality Sensitive Hashing (LSH) [66] as an algorithm for a nearest neighbour search and obtain better performance compared to SIFT or SURF approaches.

LDB

Local Difference Binary (LDB) is a highly efficient, robust and distinctive binary descriptor. The algorithm performs in, basically, three steps. First, LDB captures the internal patterns of each image patch using a set of binary tests, comparing the average intensity (I_{avg}) and first-order gradients (d_x and d_y). Second, the structure is computed at different spatial granularities. Third, the algorithm selects a subset of the bits according to their distinctiveness and concatenates them to form the final binary descriptor.

2.2.2 Image Indexing

Another important factor that affects the performance of an ABLC algorithm is its ability to efficiently retrieve past images. In this regard, ABLC approaches present some similarities with the Content Based Image Retrieval (CBIR) research field, where, given a query image, a list of similar candidate images is retrieved according to a similarity measure established between images.

A brute-force search could be considered when using global descriptors due to the simplicity of their representation. However, these descriptors are not descriptive enough and are more prone to produce false loop candidate images. Local image features are more robust, but usually several hundreds of descriptors per image are needed, increasing matching times. Therefore, in a large database of images, a brute-force search is computationally infeasible. This problem has been addressed from different points of view in the literature by means of efficient indexing schemes or using feature quantization. In this monograph, we tackle the problem from the two perspectives. Then, in this section, we briefly introduce the techniques and data structures used in future chapters for obtaining similar image candidates in an efficient way.

2.2.2.1 KD-Tree

The k-dimensional tree (kd-tree) is a space partitioning data structure for organizing points in a k-dimensional space. These structures will be used in Chap. 5 for directly indexing 128-dimensional descriptors resulting from SIFT/SURF algorithms. The kd-tree is a binary tree in which every node is a k-dimensional point. At each step of the construction of the kd-tree, one of the coordinates is selected as a splitting

Fig. 2.7 Example of a
3-dimensional tree. The first
split (red) cuts the root cell
(white) into two subcells,
each of which is then split
(green) into two subcells.
Finally, each of those four is
split (blue) into two subcells.
Image by Benjamin
Tyner/GNU GPL 2.0, via
Wikimedia Commons

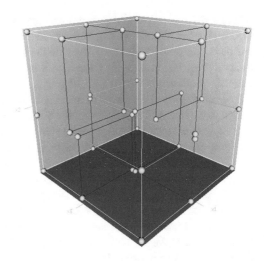

hyperplane, and then, it is used as a basis for separating the rest of the points into left and right subtrees. The point corresponding to the median value for the chosen hyperplane is usually selected as the root node. Then, the remaining points are assigned to the subtrees according to the following criterion: the items in the left subtree will have the chosen coordinate value lower than the root node, while the items in the right subtree will have the chosen coordinate value greater or equal than the root node. The process is recursively applied to the subtrees selecting another coordinate as splitting hyperplane. It finishes when all points have been allocated in the tree. An example of kd-tree with 3 dimensions is shown in Fig. 2.7.

From our point of view, the most interesting operation to be performed over a kd-tree is the Nearest Neighbour Search (NNS), which tries to find, given a query point, the closest point stored in the tree. When dealing with high-dimensional vectors, such as SIFT or SURF descriptors, the performance of a kd-tree decreases dramatically, and most of the points in the tree are required to be evaluated. This effect is known in the literature as the *curse of dimensionality*. In this case, an Approximate Nearest Neighbour Search (ANNS) method should be used instead for efficiency reasons. In this book, we tackle this problem using randomized kd-trees [67]. The idea is to build a set of independent kd-trees using the same input data. During the building process, the splitting dimensions are chosen randomly among the top five most variant dimensions at each level. The search is then performed simultaneously among all the trees and using priority queues for visiting a predefined maximum number of points.

2.2.2.2 The Bag-of-Words Model

Instead of indexing the descriptors directly, another option is to quantize the features and summarize them in clusters. In this regard, the most common method is the Bag-of-Words (BoW) algorithm. This algorithm was initially developed for text retrieval, where a BoW is a sparse vector representation of a document counting the number of occurrences of each word given a predefined vocabulary. Documents with more words in common are likely to describe the same topic. Exporting these concepts to the computer vision field [68], the idea is to treat local features as visual words and quantize them according to a set of representative features, known as *codebook* or *visual vocabulary*. This quantization is performed by mapping each descriptor of the image to the nearest image word in the dictionary. Then, the image is represented by a histogram of occurrences of each reference local feature presented in the image, reducing the total set of feature descriptors found to a vector of integers. Since some words are more discriminating than others when identifying an image, the BoW vector is normally weighted by some scoring algorithm such as the Term Frequency-Inverse Document Frequency (TF-IDF).

The most common way of generating a visual dictionary is to cluster the descriptors extracted from a set of training images using some clustering algorithm, such as k-means, where the learned centroids are considered as the reference visual words. Recently, some approaches have proposed to generate the dictionary online, avoiding this training step. In Chap. 3, we will review the main topological mapping approaches based on the BoW scheme according to these criteria.

In image retrieval, these BoW schemes are usually employed together with *inverted files*, also known as *inverted indexes*. The inverted file of a visual word is a list of the images where this word appeared. This permits to obtain rapidly image candidates as the features are processed. The topological mapping solutions proposed in this monograph take advantage of these inverted files to efficiently retrieve loop closure candidates.

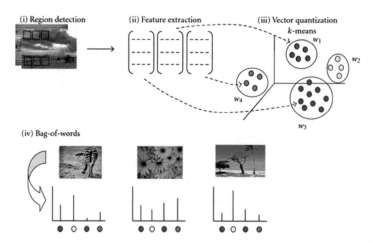

Fig. 2.8 Four main steps to obtain a BoW descriptor from an image: (i) keypoint detection; (ii) keypoint description; (iii) vocabulary construction using a clustering algorithm; and (iv) descriptor quantization from a query image to construct the final descriptor. Image by Chih-Fong Tsai/CC BY 3.0, via [69]

An example of a classic BoW scheme is shown in Fig. 2.8. A clustering technique (k-means in the figure) is performed against the set of training descriptors in order to obtain the visual vocabulary, which corresponds to the centroid of each obtained cluster. When a new image needs to be processed, its extracted descriptors are associated to the closest visual word in the dictionary and then, a histogram of occurrences of each visual word is computed. Then, a distance metric between histograms, such as the cosine distance, could be used to estimate the similarity between the two image representations.

References

1. Busquets, D.: A multi-agent approach to qualitative navigation in robotics. Ph.D. thesis, Universitat Politècnica de Catalunya (2003)
2. Dudek, G., Jenkin, M.: Computational Principles of Mobile Robotics. Cambridge University Press (2010)
3. Korrapati, H.: Loop closure for topological mapping and navigation with omnidirectional images. Ph.D. thesis, Université Blaise Pascal-Clermont-Ferrand II (2013)
4. Wu, J., Rehg, J.M.: CENTRIST: a visual descriptor for scene categorization. IEEE Trans. Pattern Anal. Mach. Intell. **33**(8), 1489–1501 (2011)
5. Bosch, A., Zisserman, A., Munoz, X.: Representing shape with a spatial pyramid kernel. ACM Int. Conf. Image Video Ret. 401–408 (2007)
6. Fazl-Ersi, E., Tsotsos, J.K.: Histogram of oriented uniform patterns for robust place recognition and categorization. Int. J. Rob. Res. **31**(4), 468–483 (2012)
7. Zhou, L., Zhou, Z., Hu, D.: Scene classification using a multi-resolution bag-of-features model. Patt. Recog. **46**(1), 424–433 (2013)
8. Dalal, N., Triggs, B.: Histograms of oriented gradients for human detection. IEEE Conf. Comput. Vision Pattern Recog. 886–893 (2005)
9. Winters, N., Gaspar, J., Lacey, G., Santos-Victor, J.: Omni-directional vision for robot navigation. In: IEEE Workshop on Omnidirectional Vision, pp. 21–28 (2000)
10. Gaspar, J., Winters, N., Santos-Victor, J.: Vision-based navigation and environmental representations with an omnidirectional camera. IEEE Trans. Robot. Autom. **16**(6), 890–898 (2000)
11. Ulrich, I., Nourbakhsh, I.: Appearance-based place recognition for topological localization. IEEE Int. Conf. Robot. Autom. **2**, 1023–1029 (2000)
12. Kosecka, J., Zhou, L., Barber, P., Duric, Z.: Qualitative image based localization in indoors environments. In: IEEE Conf. Comput. Vision Pattern Recog. **2**, pp. II–3–II–8 (2003)
13. Bradley, D., Patel, R., Vandapel, N., Thayer, S.: Real-time image-based topological localization in large outdoor environments. IEEE/RSJ Int. Conf. Intell. Robots Syst. 3670–3677 (2005)
14. Weiss, C., Masselli, A.: Fast outdoor robot localization using integral invariants. IEEE Int. Conf. Comput. Vision, 1–10 (2007)
15. Wang, J., Zha, H., Cipolla, R.: Efficient topological localization using orientation adjacency coherence histograms. Int. Conf. Pattern Recog. 271–274 (2006)
16. Pronobis, A., Caputo, B., Jensfelt, P., Christensen, H.: A discriminative approach to robust visual place recognition. In: IEEE/RSJ Int. Conf. Intell. Robots Syst. 3829–3836 (2006)
17. Oliva, A., Torralba, A.: Modeling the shape of the scene: a holistic representation of the spatial envelope. Int. J. Comput. Vision **42**(3), 145–175 (2001)
18. Murillo, A.C., Campos, P., Kosecka, J., Guerrero, J.: Gist vocabularies in omnidirectional images for appearance based mapping and localization. In: Workshop on Omnidirectional Vision, Camera Networks and Non-classical Cameras (RSS) (2010)
19. Sunderhauf, N., Protzel, P.: BRIEF-gist - closing the loop by simple means. IEEE/RSJ Int. Conf. Intell. Robots Syst. 1234–1241 (2011)

20. Chapoulie, A., Rives, P., Filliat, D.: Appearance-based segmentation of indoors and outdoors sequences of spherical views. IEEE Int. Conf. Robot. Autom. 1946–1951 (2013)
21. Lamon, P., Nourbakhsh, I., Jensen, B., Siegwart, R.: Deriving and matching image fingerprint sequences for mobile robot localization. IEEE Int. Conf. Robot. Autom. **2**, 1609–1614 (2001)
22. Liu, M., Scaramuzza, D., Pradalier, C., Siegwart, R., Chen, Q.: Scene recognition with omnidirectional vision for topological map using lightweight adaptive descriptors. IEEE/RSJ Int. Conf. Intell. Robots Syst. 116–121 (2009)
23. Liu, M., Siegwart, R.: DP-FACT: towards topological mapping and scene recognition with color for omnidirectional camera. IEEE Int. Conf. Robot. Autom. 3503–3508 (2012)
24. Menegatti, E., Maeda, T., Ishiguro, H.: Image-based memory for robot navigation using properties of omnidirectional images. Rob. Auton. Syst. **47**(4), 251–267 (2004)
25. Menegatti, E., Zoccarato, M., Pagello, E., Ishiguro, H.: Image-based monte carlo localisation with omnidirectional images. Rob. Auton. Syst. **48**(1), 17–30 (2004)
26. Prasser, D., Wyeth, G.: Probabilistic visual recognition of artificial landmarks for simultaneous localization and mapping. IEEE Int. Conf. Robot. Autom. **1**, 1291–1296 (2003)
27. Milford, M., Wyeth, G.: Mapping a suburb with a single camera using a biologically inspired slam system. IEEE Trans. Robot. **24**(5), 1038–1053 (2008)
28. Milford, M., Wyeth, G.: SeqSLAM: Visual route-based navigation for sunny summer days and stormy winter nights. IEEE Int. Conf. Robot. Autom. 1643–1649 (2012)
29. Lui, W.L.D., Jarvis, R.: A pure vision-based approach to topological slam. IEEE/RSJ Int. Conf. Intell. Robots Syst. 3784–3791 (2010)
30. Lui, W.L.D., Jarvis, R.: A pure vision-based topological slam system. Int. J. Robt. Res. **31**(4), 403–428 (2012)
31. Badino, H., Huber, D., Kanade, T.: Real-time topometric localization. IEEE Int. Conf. Robot. Autom. 1635–1642 (2012)
32. Lategahn, H., Beck, J., Kitt, B., Stiller, C.: How to learn an illumination robust image feature for place recognition. Intell. Vehic. Symp. 285–291 (2013)
33. Nourani-Vatani, N., Borges, P., Roberts, J., Srinivasan, M.: On the use of optical flow for scene change detection and description. J. Intell. Robot. Syst. **74**(3), 817–846 (2014)
34. Tuytelaars, T., Mikolajczyk, K.: Local Invariant Feature Detectors: A Survey. Found. Trends® Comput. Gr. Vis. **3**(3), 177–280 (2007)
35. Schmidt, A., Kraft, M., Kasinski, A.: An evaluation of image feature detectors and descriptors for robot navigation. ICCVG, Computer Vision and Graphic. Lecture Notes in Computer Science, pp. 251–259. Springer, Berlin (2010)
36. Miksik, O., Mikolajczyk, K.: Evaluation of local detectors and descriptors for fast feature matching. Int. Conf. Pattern Recog. 2681–2684 (2012)
37. Mikolajczyk, K., Schmid, C.: A performance evaluation of local descriptors. IEEE Trans. Pattern Anal. Mach. Intell. **27**(10), 1615–1630 (2005)
38. Harris, C., Stephens, M.: A Combined Corner and Edge Detector. In: Alvey Vision Conference, pp. 147–151 (1988)
39. Shi, J., Tomasi, C.: Good features to track. IEEE Conf. Comput. Vision Pattern Recog. 593–600 (1994)
40. Smith, S., Brady, M.: SUSAN - a new approach to low level image processing. Int. J. Comput. Vision **23**(1), 45–78 (1997)
41. Rosten, E., Drummond, T.: Machine learning for high-speed corner detection. Eur. Conf. Comput. Vision, 430–443 (2006)
42. Rosten, E., Porter, R., Drummond, T.: Faster and better: a machine learning approach to corner detection. IEEE Trans. Pattern Anal. Mach. Intell. **32**(1), 105–19 (2010)
43. Rublee, E., Rabaud, V., Konolige, K., Bradski, G.: ORB: an efficient alternative to sift or surf. IEEE Int. Conf. Comput. Vision **95**, 2564–2571 (2011)
44. Mair, E., Hager, G.D., Burschka, D., Suppa, M., Hirzinger, G.: Adaptive and generic corner detection based on the accelerated segment test. European Conference on Computer Vision. Lecture Notes in Computer Science, vol. 6312, pp. 183–196. Springer, Berlin (2010)

45. Leutenegger, S., Chli, M., Siegwart, R.: BRISK: Binary robust invariant scalable keypoints. IEEE Int. Conf. Comput. Vision, 2548–2555 (2011)
46. Lowe, D.G.: Distinctive image features from scale-invariant keypoints. Int. J. Comput. Vision **60**(2), 91–110 (2004)
47. Bay, H., Tuytelaars, T., Van Gool, L.: SURF: speeded up robust features. European Conference on Computer Vision. Lecture Notes in Computer Science, vol. 3951, pp. 404–417. Springer, Berlin (2006)
48. Agrawal, M., Konolige, K., Blas, M.R.: CenSurE: center surround extremas for realtime feature detection and matching. European Conference on Computer Vision, vol. 5305, pp. 102–115. Springer, Berlin (2008)
49. Konolige, K., Bowman, J., Chen, J., Mihelich, P., Calonder, M., Lepetit, V., Fua, P.: View-based maps. Int. J. Robt. Res. **29**(8), 941–957 (2010)
50. Ebrahimi, M., Mayol-Cuevas, W.: SUSurE: Speeded up surround extrema feature detector and descriptor for realtime applications. IEEE Conf. Comput. Vision Pattern Recog. 9–14 (2009)
51. Alcantarilla, P.F., Bartoli, A., Davison, A.J.: KAZE features. European Conference on Computer Vision, pp. 214–227. Springer, Berlin (2012)
52. Alcantarilla, P.F., Nuevo, J., Bartoli, A.: Fast explicit diffusion for accelerated features in nonlinear scale spaces. In: British Machine Vision Conference (BMVC) (2013)
53. Morel, J.M., Yu, G.: ASIFT: a new framework for fully affine invariant image comparison. SIAM J. Imaging Sci. **2**(2), 438–469 (2009)
54. Matas, J., Chum, O., Urban, M., Pajdla, T.: Robust wide baseline stereo from maximally stable extremal regions. In: British Machine Vision Conference (BMVC), pp. 1–10 (2002)
55. Ke, Y., Sukthankar, R.: PCA-SIFT: A more distinctive representation for local image descriptors. IEEE Conf. Comput. Vision Pattern Recog. 506–513 (2004)
56. Andreasson, H., Duckett, T.: Topological localization for mobile robots using omnidirectional vision and local features. IFAC Symp. Intell. Auton, Vehic (2008)
57. Tola, E., Lepetit, V., Fua, P.: DAISY: an efficient dense descriptor applied to wide baseline stereo. IEEE Trans. Pattern Anal. Mach. Intell. **32**(5), 815–830 (2010)
58. Sarfraz, M.S., Hellwich, O.: Head Pose Estimation in Face Recognition Across Pose Scenarios. In: International Conference on Computer Vision Theory and Applications, pp. 235–242 (2008)
59. Calonder, M., Lepetit, V., Strecha, C., Fua, P.: BRIEF : binary robust independent elementary features. European Conference on Computer Vision. Lecture Notes in Computer Science, vol. 6314, pp. 778–792. Springer, Berlin (2010)
60. Alahi, A., Ortiz, R., Vandergheynst, P.: FREAK : fast retina keypoint. IEEE Conf. Comput. Vision Pattern Recog. 510–517 (2012)
61. Trzcinski, T., Lepetit, V.: Efficient discriminative projections for compact binary descriptors. European Conference on Computer Vision. Lecture Notes in Computer Science, vol. 7572, pp. 228–242 (2012)
62. Strecha, C., Bronstein, A.M., Bronstein, M.M., Fua, P.: LDAHash: Improved matching with smaller descriptors. IEEE Trans. Pattern Anal. Mach. Intell. **34**(1) (2012)
63. Trzcinski, T., Christoudias, C., Fua, P., Lepetit, V.: Boosting binary keypoint descriptors. IEEE Conf. Comput. Vision Pattern Recog. 2874–2881 (2013)
64. Yang, X., Cheng, K.T.: Local difference binary for ultrafast and distinctive feature description. IEEE Trans. Pattern Anal. Mach. Intell. **36**(1), 188–94 (2014)
65. Geng, L.C., Jodoin, P.M., Su, S.Z., Li, S.Z.: CBDF: compressed binary discriminative feature. Neurocomputing **184**, 43–54 (2015)
66. Gionis, A., Indyk, P., Motwani, R.: Similarity search in high dimensions via hashing. In: International Conference on Very Large Data Bases, pp. 518–529 (1999)
67. Silpa-Anan, C., Hartley, R.: Optimised kd-trees for fast image descriptor matching. IEEE Conf. Comput. Vision Pattern Recog. 1–8 (2008)
68. Sivic, J., Zisserman, A.: video google: a text retrieval approach to object matching in videos. IEEE Int. Conf. Comput. Vision, 1470–1477 (2003)
69. Tsai, C.F.: Bag-of-words representation in image annotation: a review. ISRN Artif. Intell. **2012** (2012)

Chapter 3
Literature Review

Abstract This chapter reviews the main approaches published during the last years with regard to topological mapping and localization by visual means. We classify the different solutions according to the method used to visually describe an image, given the fact that the quality of the resulting map strongly relies on this aspect. Three fundamental categories are distinguished: approaches based on global descriptors, approaches based on local features and approaches based on Bag-Of-Words (BoW) schemes. We also consider different combinations of these methods.

3.1 Overview

In this chapter we review the main approaches published in the last fifteen years with regard to topological mapping and localization by visual means. In the related litera- ture, one can find similar surveys, although they rather focus more on navigation [1] and on visual SLAM [2]. In this review we will mostly consider approaches dealing with topological maps, although we will also take into account hybrid solutions that somehow consider the topology of the environment. Other possibly related problem is that of pose-graph SLAM. Notice that algorithms such as Olson [3], TreeMap [4], Square Root SAM [5], iSAM [6], TORO [7], Sparse Pose Adjustment [8], iSAM2 [9] or g2o [10] could take as input a topological map. However, pose-graph SLAM nodes usually represent poses reached by the agent, and not distinctive places of the envi- ronment. Besides, the position in pose-graph SLAM is a metric position of the vehicle and not a qualitative estimation in a discrete model of the appearance of the world. Because of those reasons, we will consider this class of mapping algorithms out of the scope of this chapter.

Loop closure detection is an important component in topological schemes. When using vision as a source, this problem is usually solved comparing images directly, resulting into appearance-based approaches. In this regard, a related research field is *scene categorization* or *visual place categorization* (VPC) [11]. The main goal of this area is to find the class of a place in a rough manner. For instance, given the current image, the objective is to conclude that the current place is a kitchen. Some authors create topological maps using these frameworks, forming a graph of known

© Springer International Publishing AG 2018   27
E. Garcia-Fidalgo and A. Ortiz, *Methods for Appearance-based Loop Closure Detection*, Springer Tracts in Advanced Robotics 122, https://doi.org/10.1007/978-3-319-75993-7_3

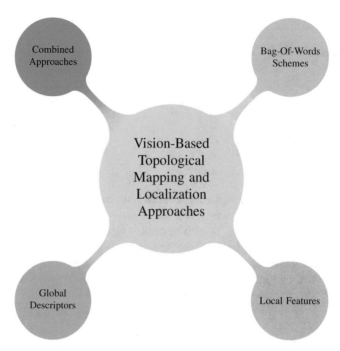

Fig. 3.1 Taxonomy for classifying vision-based topological schemes according to their image representation method

places. However, VPC can be considered as a different research line and these works are also out of the scope of this chapter.

In order to perform mapping and localization tasks using vision, it is necessary to describe the acquired images and be able to compare these descriptions. Consequently, the quality of the map and the posterior localization will directly rely on the method used for visually describing the different environment locations. For this reason, we classify the different approaches according to the description method employed as: approaches based on global descriptors, approaches based on local features and approaches based on Bag-Of-Words (BoW) schemes. We also identify that these methods can be combined. See Fig. 3.1 for a graphical description of this classification.

Note that BoW schemes, where the local features extracted from an image are quantized according to a set of representative visual words, are mainly used in combination with an inverted file to index visual information in an efficient way for fast image retrieval, and could be regarded as a subcategory within the local features approaches. Another possibility is to consider that, in these approaches, the image representation changes from the set of local features to a histogram of occurrences of each visual word in the image, reducing the descriptor to a vector of integers. In this chapter, and to make it more understandable, we have decided to adopt the second view, keep the BoW category and, hence, classify the BoW-related papers apart from the local features-related papers.

Given the taxonomy of the problem, the rest of the chapter is organized as follows: Sect. 3.2 enumerates fundamental works based on global descriptors, approaches based on local features are presented in Sect. 3.3, Sect. 3.4 introduces main solutions built under BoW schemes, Sect. 3.5 enumerates principal works that represent the image as a combination of the other ones and Sect. 3.6 concludes the chapter.

3.2 Methods Based on Global Descriptors

Many authors have proposed different solutions for topological mapping and localization using global image representations, which are summarized in Table 3.1. This table indicates, for each solution, the imaging configuration adopted, whether the resulting map is a pure topological map or otherwise is a hybrid representation, the intended tasks, the environments where the approach was assessed and the image descriptor used. For further information about global descriptors, the reader is referred to Sect. 2.2.1.1.

3.2.1 Histograms

Histograms provide a compact way of representing an image and have been used for topological mapping and localization in different forms. An example of that is the work of Ulrich and Nourbakhsh [14]. They proposed a topological localization method based on appearance. Each image is represented by six one-dimensional colour histograms, three extracted from the HLS colour space and other three extracted from the RGB colour space. Given a query image, they retrieved reference images from the map using a nearest neighbour learning scheme in their topological map. The Jeffrey divergence was used as a distance measure between two histograms. They assessed their system in several environments, obtaining at least 87.5% of correctly classified images in all of them. Werner et al. [15] also employed colour histograms combined with a Bayes filter for providing a topological SLAM solution. They used the Hausdorff distance to compare the topological map and the visual observations received by the robotic platform. They argued that colour histograms are not distinctive enough, and that the Bayes filter helps to disambiguate places with similar appearance.

Kosecka et al. [16] proposed a navigation strategy using gradient orientation histograms as image descriptor. In an exploration phase, a topological map was built by comparing successive frame descriptors. For each node, a set of representative views was computed using Learning Vector Quantization (LVQ). During the navigation, the current frame's histogram was extracted and compared with each node representatives using the Euclidean distance to determine the most similar location.

Table 3.1 Summary of topological mapping and localization solutions based on global image descriptors

References	Camera	Map	Tasks	Environment	Descriptor
Winters [12]	Omnidir	Topo	Map + Loc	Indoors	PCA
Gaspar [13]	Omnidir	Topo	Map + Loc	Indoors	PCA
Ulrich [14]	Omnidir	Topo	Map + Loc	In + Out	Colour Hist.
Werner [15]	Omnidir	Topo	SLAM	Indoors	Colour Hist.
Kosecka [16]	Mono	Topo	Map + Loc	Indoors	Gradient Orien. Hist.
Bradley [17]	Mono	Topo	Map + Loc	Outdoors	WGOH
Weiss [18]	Mono	Topo	Map + Loc	Outdoors	WGII
Wang [19]	Mono	Topo	Map + Loc	In + Out	OACH
Pronobis [20]	Mono	Topo	Loc	Indoors	Receptive Field Hist.
Singh [21]	Omnidir	Topo	Map + Loc	Outdoors	Gist
Murillo [22]	Omnidir	Hybrid	Map + Loc	In + Out	Omni-Gist
Rituerto [23]	Omnidir	Topo	Mapping	Indoors	Omni-Gist
Sunderhauf [24]	Mono	Topo	SLAM	Outdoors	BRIEF-Gist
Arroyo [25]	Omnidir	Topo	Map + Loc	Outdoors	LDB
Arroyo [26]	Stereo	Topo	Map + Loc	Outdoors	D-LDB
Liu [27]	Mono	Topo	SLAM	Outdoors	Gist
Chapoulie [28]	Sphere	Topo	Map + Loc	In + Out	Gist
Chapoulie [29]	Sphere	Topo	Map + Loc	In + Out	Spherical Harmonics
Lamon [30]	Omnidir	Topo	Loc	Indoors	Fingerprints
Tapus [31, 32]	Omnidir	Topo	Map + Loc	Indoors	Fingerprints
Liu [33]	Omnidir	Topo	Mapping	Indoors	FACT
Liu [34]	Omnidir	Topo	Mapping	Indoors	DP-FACT
Menegatti [35, 36]	Omnidir	Topo	Map + Loc	Indoors	Fourier Signatures
Paya [37]	Omnidir	Topo	Map + Loc	Indoors	Fourier Signatures
Ranganathan [38]	Omnidir	Topo	Mapping	Indoors	Fourier Signatures
Milford [39]	Mono	Hybrid	SLAM	Indoors	Colour Segmentation
Prasser [40]	Omnidir	Hybrid	SLAM	Outdoors	Colour Hist.
Milford [41]	Mono	Hybrid	SLAM	Outdoors	Scan Intensity Prof.
Glover [42]	Mono	Hybrid	SLAM	Outdoors	Scan Intensity Prof.
Lui [43, 44]	Omnidir	Hybrid	SLAM	In + Out	2D Haar Wavelet Dec.
Badino [45]	Mono	Hybrid	Map + Loc	Outdoors	WI-SURF
Xu [46]	Mono	Hybrid	Map + Loc	Outdoors	WI-SURF
Lategahn [47]	Mono	Hybrid	SLAM	Outdoors	DIRD
Nourani [48]	Mono	Topo	Map + Loc	In + Out	OFM/OFSC
Milford [49–51]	Mono	Topo	SLAM	Outdoors	Normalized Patches
Pepperell [52]	Mono	Topo	SLAM	Outdoors	Normalized Patches
Wu [53]	Mono	Topo	Map + Loc	Outdoors	Binarized Patches

Inspired by Kosecka's work, Bradley et al. [17] introduced a topological localization approach in large outdoor environments using Weighted Gradient Orientation Histogram (WGOH) features. These features were computed partitioning the image into a grid, and extracting an 8-bin histogram of the gradient orientations for each part of the grid, weighted by the magnitude of the gradient at each point and the distance from the centre to the region. A WGOH descriptor was formed concatenating each histogram and normalizing it to the unit length. In order to avoid a dependence of the feature vector to any particular component, values higher than 0.2 were capped to 0.2 and the final descriptor was re-normalized again. Their experiments covered over 100,000 images and 67 km of traverse with a high success. Similarly, Weiss et al. [18] also split each image into a grid, but computing an 8×8 histogram of integral invariants using two relational kernels. These integral invariant features are features which are invariant to some Euclidean motions, such as rotations or translations. The main idea is to apply all possible transformations to each sub-image and obtain and averaged version of these image transformations. They called this approach Weighted Grid Integral Invariant (WGII) features. These features were combined with a particle filter for outdoor mobile robot localization. Wang et al. [19] introduced Orientation Adjacency Coherence Histograms (OACH) to solve the coarse part of a topological localization process. OACH is an extension of the traditional gradient orientation histograms where two Orientation Adjacency Histograms (OAH) are computed respectively in the edge and corner regions of the image according to the Harris detector response and concatenated to form the final descriptor. In an OAH, the gradient orientations of the center pixel's 4-neighbourhood are accumulated and then normalized by the number of centre pixels of each orientation. The Jeffrey divergence between OACH descriptors was used to compare the images in the framework.

Pronobis et al. [20] showed that receptive field responses summarized into histograms can be used for place recognition. In a training phase, several histograms were acquired from the environment and used to train Support Vector Machines (SVM) as classifiers which served as a basis of a topological localization process.

3.2.2 The Gist Descriptor

Recently, several approaches have proposed to use the Gist global descriptor [54]. Initially developed for scene recognition, it is based on the observation that humans are able to classify images at a single glance under certain conditions. Their authors concluded that humans are receptive to what they called the *spatial envelope* of the scene, defined as a set of perceptual properties related to the shape of the space. They demonstrated that this spatial envelope is closely correlated with second-order statistics (Discriminant Spectral Template) and with the spatial arrangement of structures in the scene (Windowed Discriminant Spectral Template). A bank of filters (such as Gabor filters [55]) can be used to infer a global descriptor of the scene. Principal

Component Analysis (PCA) can be also used in order to reduce the final dimension of the descriptor.

Singh and Kosecka [21] computed a Gist descriptor for panoramas applying the algorithm to each of the four views that the omnidirectional image consisted of. They introduced a novel similarity measure between image panoramas for these descriptors and evaluated its efficiency for loop closure detection in urban environments. Murillo et al. [22] extended this proposal and introduced *omni-gist*, an adapted version of the descriptor to be used with omnidirectional images extracted from catadioptric cameras, instead of multi-camera systems. They improved the similarity measure for these descriptors and proposed a hierarchical topological localization and map building algorithm based on them. In a more recent work [23], omni-gist was used in a semantic labelling process for building indoor topological maps. The images were classified as *places* or *transitions*, which corresponds to, respectively, the nodes and the edges of the topological map. This place classification module was integrated with a Hidden Markov Model (HMM) to ensure the temporal consistency.

Liu and Zhang [27] employed PCA to reduce the dimensionality of a Gist descriptor for improving the efficiency and the discriminative power of the descriptor. Then, they presented a particle filter for detecting loop closures in a SLAM system. These descriptors were taken into account in the update step of the filter. As a result, they showed that a high recall can be obtained at 100% precision with only a few particles.

Chapoulie et al. [28] presented an approach for segmenting the environment into topological places using spherical images. This segmentation approach was based on detecting changes in the environment and an adapted version of Gist for spherical images. In a more recent work [29], they argued that Gist is not well adapted to represent this kind of images because the sphere spatial periodicity is partially lost. Then, they introduced a new global image representation based on spherical harmonics adapted for spherical views.

Finally, motivated by the success of Gist and the BRIEF binary descriptor [56], Sunderhauf and Protzel [24] adapted the latter to be used as a global descriptor, introducing the BRIEF-Gist descriptor. The implementation is very straightforward: the image is downsampled to the size of a patch and a BRIEF descriptor is computed from its centre. Other possible implementation consists in partitioning the image into a grid, compute the BRIEF descriptor for each patch and concatenate them to form the final descriptor. They used this simple descriptor for loop closing in a SLAM system that can be used in a large-scale scenario, as is shown in their experiments. As a main drawback, BRIEF-Gist is not able to detect bidirectional loops. In this regard, Arroyo et al. [25] introduced an algorithm called Able for Binary-appearance Loop-closure Evaluation applied to Panoramas (ABLE-P) which can detect these cases. They divided each panorama in sub-panoramas and extracted an LDB binary descriptor for each of them [57]. The final image descriptor is created concatenating the different LDB strings. The loop closures are then found correlating the descriptors of the different panoramas using the Hamming distance. In a more recent work [26], they updated their algorithm to be used with a monocular or a stereo camera (ABLE-S) and added disparity information to the LDB descriptor, generating the D-LDB descriptor, which is also used for detecting loop closures.

3.2.3 Vertical Regions

Extracting vertical lines in order to define globally omnidirectional images has also been used for topological mapping and localization, specially for indoor environments because of the nature of their structures. In this regard, Lamon et al. [30] presented the concept of *fingerprints* of places. A fingerprint is a circular list of features extracted using different algorithms. In their case, they used two detectors: a vertical edge detector based on histograms and a colour patch detector. They also presented an algorithm for matching these sequences of features based on a minimum energy algorithm, and employed this framework for global localization. Tapus et al. [31] demonstrated that this fingerprint representation combined with an uncertainty model of the features can improve the localization results. After this work, Tapus and Siegwart [32] expanded the fingerprint concept incorporating information from a laser range finder in an incremental topological mapping approach for multi-room indoor environments.

Liu et al. [33] introduced the Fast Adaptive Color Tags (FACT) descriptor, employed for a topological mapping approach. It is based on the fact that, in indoor environments, the important vertical edges (windows, columns, and so on) naturally divide the indoor environment into several meaningful cuts. For each cut, the average colour value in the U-V space is computed. This U-V average value and the width of the region form a region descriptor called *tag*. A scene descriptor is formed concatenating each region descriptor in a vector. Scene matching between new scenes and existing nodes was performed computing the 2D Euclidean distance between colour descriptors, and recursively comparing the widths of the regions according to an empirically determined inequality. In order to take into the account the main drawbacks that this solution presented, they improved their descriptor publishing another version called DP-FACT [34], where a Dirichlet Process Mixture Model is used to combine colour and geometry features extracted from omnidirectional images.

3.2.4 Discrete Fourier Transform

Several authors have proposed to use the Discrete Fourier Transform (DFT) as a global image representation method. Menegatti et al. [35] unwarped omnidirectional images over a panoramic cylinder. These panoramic cylinders were expanded row by row into their Fourier series. An image was represented by the first 15 Fourier coefficients i.e. the 15 lowest frequency components, reducing the storage needs for each reference view. The set of these selected coefficients was called by their authors as *Fourier signatures*. They also proposed a method for an automatic organization of a set of reference images obtained in an exploration phase into a *visual memory* and a navigation approach using this framework. To overcome the perceptual aliasing problem that the original approach presented, in a following work [36], they improved their localization system fusing this image representation with a particle filter. Based on these works, Paya et al. [37] contributed with an incremental mapping process, creating the map while the robot is traversing the environment and Ranganathan et al. [38] introduced the concept of Probabilistic Topological Maps (PTM), where a particle filter was employed for approximating the posterior distribution over the possible topologies given the available sensor measurements and an odometry source.

3.2.5   Biologically-Inspired Approaches

Biologically-inspired solutions try to emulate the information processing methods and problem resolution abilities of the biological systems, simulating the behaviour of living organisms. Several topological mapping and localization solutions fall under this subcategory.

Gaspar et al. [13] mapped an indoor environment emulating the vision-based navigation capabilities of insects using an omnidirectional camera. The images of the topological map were encoded as a manifold in a low-dimensional eigenspace obtained from PCA. In an offline phase, they created a representation of the environment resulting into a topological map, which was later used to navigate using a visual following approach.

Milford et al. [39] introduced RatSLAM, a single-camera SLAM system derived from models of the hippocampal complex in rodents. According to the authors, the operation of these models appears to be related with some topological and metric properties to its advantage, so it can be considered as a hybrid approach. The environment representation was built using a competitive attractor network structure called *pose cells*, which was used to concurrently represent the belief about the location and orientation of the robot. The system performed a colour segmentation process [58] to detect some coloured cylinders spread around the experimental area in order to update these pose cells. This approach was later adapted by Prasser et al. [40] to be used in outdoor environments and using an omnidirectional camera as a main input sensor. Images were described using histograms of the hue and saturation colour bands and compared using the χ^2 statistic. Later, Milford and Wyeth [41] mapped a path of 66 km along an entire suburb using RatSLAM, showing that it can be used in a long-term operation. A scanline intensity profile is employed as image descriptor, which is a one-dimensional vector formed by summing the intensity values in each pixel column, and then normalizing the final vector. Glover et al. [42] combined RatSLAM with other approaches in order to address the challenging problem of producing coherent maps across several times of the day.

3.2.6   Other Approaches

Winters et al. [12] utilized an omnidirectional camera to create a topological map from the environment during a training phase. Nodes were sets of images with common properties, and links were sequences of consecutive views between two nodes. The large image set obtained was compressed using PCA, resulting in a low-dimensional eigenspace from which the robot could determine its global topological position using an appearance-based method.

Badino [45] presented an outdoor localization approach based in a descriptor called Whole Image SURF (WI-SURF), where a Speeded Up Robust Feature (SURF) descriptor for the entire image is computed according to [59]. Each node of the map is associated with the GPS coordinates where it was acquired, and a Bayesian filter is used to compute the probability of being in each discrete place of the map. They reported successful results for long-term localization experiments, concluding its validity for solving the global localization problem. In a more recent work [46], they presented an algorithm for localizing a vehicle on an arbitrary road network.

Lategahn et al. [47] studied how to generate robust descriptors for environments under severe lighting changes. They proposed to use building blocks which can be used to construct millions of descriptors. In that work, an evaluation function to evaluate the performance of these descriptors was presented, as well as a search algorithm for them. Results for loop closure detection were also presented. The experiments were carried on using the best combination of these building blocks found and was called *Dird is an Illumination Robust Descriptor* (DIRD).

A complete loop closing system for autonomous mobile robots was proposed by Lui and Jarvis [43, 44], where omnidirectional images was described employing a GPU-based 2D Haar Wavelet decomposition. These images are used to create a database of signatures. A relaxation algorithm is executed to adjust the topology each time the vehicle revisits a previously seen place.

Nourani–Vatani et al. [48] proposed to use optical flow information to detect changes in the environment, using the Optical Flow Moment (OFM) and the Optical Flow Shape Context (OFSC) descriptors. Then, statistical attributes from the flow were extracted in order to define each location. Once a database of nodes was generated, where a node was defined as a detected scene change, the most likely location was obtained using the Mahalanobis and χ^2 distances. They assessed their approach in indoor and outdoor environments, showing that it could be used in several kinds of scenarios.

In a more recent research line, Milford and Wyeth presented SeqSLAM [49], where instead of searching for a single previously seen image given the current frame, they performed the localization process recognizing coherent sequences of local consecutive images. They showed that this approach could be used for visual navigation under weather or season changes. They employed normalized patches in a cropped version of the original image, and Sum of Absolute Differences (SAD) to compare these patches. They have also showed that route recognition can be accomplished even with a few bits per image [50] and they studied the effect of the length of the sequences onto the SeqSLAM algorithm performance [51]. An evolution of the SeqSLAM algorithm called Sequence Matching Across Route Traversals (SMART) has been recently proposed in [52], which improves its general applicability by integrating self-motion information to form spatially consistent sequences, and new image matching techniques to handle greater perceptual change and variations in translational pose.

Wu et al. [53] presented a loop closure detection method which uses an extremely simple image representation. Images are smoothed using a Gaussian kernel, and then resized to a small patch. The Otsu's method is then employed to binarize the image, producing a binary code of a few hundred bits. The mutual information for the image pairs is used as a similarity measure. According to their results, they are able to detect loop closures in a map of 20 million key locations.

3.3   Methods Based on Local Features

Several authors, as shown in Table 3.2, have used local image features to perform topological mapping and localization tasks, specially since the release of the Lowe's Scale-Invariant Feature Transform (SIFT) algorithm. Kosecka and Yang [60, 61] used SIFT features for describing images in indoor environments and performed a global localization process based on a simple voting scheme. In order to overcome the problems resulting from dynamic changes in the environment, they proposed to incorporate additional knowledge about neighbourhood relationships between individual locations using a Hidden Markov Model. The likelihood function was based on the number of correspondences between the current image and past locations. Following this work, in [62] they presented a feature selection strategy in order to reduce the number of keypoints per location. This strategy was carried on measuring the discriminability of the individual features to describe each topological location. Zhang [63] also presented a method for selecting a subset of visual features from an image called Bag-of-Raw-Features (BoRF). The features were selected according to the scale where they were found. A location was represented by the set of features that can be matched consecutively in several images, applying a keyframe selection policy based on their previous work [100]. The main problem that BoRF presents was that the number of features to manage increases while new images were added, and a linear search for matching became intractable. This drawback was overcome in [64] by indexing features through kd-tree structures.

Using the idea of maintaining only persistent features, several authors have proposed various solutions to the community. Rybski et al. [65] used Kanade–Lucas–Tomasi (KLT) feature tracker for matching persistent features in a sequence of omnidirectional images and constructed a topological map incrementally. He et al. [66] proposed to use manifold constraints to find representative feature prototypes, which are useful to represent any image within the environment in an efficient manner. Sabatta [67] introduced a mapping and localization algorithm that exploits the persistence of SIFT features within consecutive omnidirectional images to improve data association. He also modified the SIFT algorithm in order to include colour information in the descriptor. More recently, Johns and Yang [68] introduced an approach where the map is composed by a set of landmarks detected across multiple images, spanning the continuous space between nodal images. Given a query image, matches

Table 3.2 Summary of topological mapping and localization solutions based on local features

References	Camera	Map	Tasks	Environment	Feature
Kosecka [60–62]	Mono	Topo	Map + Loc	Indoors	SIFT
Zhang [63]	Mono	Topo	Map + Loc	Indoors	SIFT
Zhang [64]	Mono	Topo	SLAM	Indoors	SIFT
Rybski [65]	Omnidir	Topo	Map + Loc	Indoors	KLT
He [66]	Mono	Topo	Map + Loc	Outdoors	SIFT
Sabatta [67]	Omnidir	Topo	Map + Loc	Indoors	SIFT
Johns [68]	Mono	Topo	Map + Loc	Indoors	SIFT
Kawewong [69, 70]	Omnidir	Topo	SLAM	In + Out	PIRF (SIFT)
Tongprasit [71]	Omnidir	Topo	SLAM	In + Out	PIRF (SURF)
Morioka [72]	Omnidir	Hybrid	SLAM	Indoors	3D-PIRF (SURF)
Andreasson [73]	Omnidir	Topo	Map + Loc	Indoors	KLT/M-SIFT
Valgren [74]	Omnidir	Topo	Mapping	Indoors	KLT/M-SIFT
Valgren [75]	Omnidir	Topo	Mapping	In + Out	SIFT
Valgren [76]	Omnidir	Topo	Loc	Outdoors	SIFT/SURF
Ascani [77]	Omnidir	Topo	Loc	In + Out	SIFT/SURF
Anati [78]	Omnidir	Topo	Map + Loc	In + Out	SIFT
Zivkovic [79]	Omnidir	Hybrid	Map + Loc	Indoors	SIFT
Booij [80]	Omnidir	Hybrid	Map + Loc	Indoors	SIFT
Booij [81]	Omnidir	Hybrid	Map + Loc	In + Out	SIFT
Dayoub [82]	Omnidir	Hybrid	Map + Loc	Indoors	SURF
Blanco [83, 84]	Stereo	Hybrid	SLAM	Indoors	SIFT
Tully [85]	Omnidir	Hybrid	Map + Loc	Indoors	SIFT
Tully [86]	Omnidir	Hybrid	SLAM	Indoors	SIFT
Segvic [87]	Mono	Hybrid	Map + Loc	Outdoors	SIFT/Harris/MSER
Ramisa [88]	Omnidir	Topo	Map + Loc	Indoors	MSER/SIFT/GLOH
Badino [89]	Mono	Hybrid	Map + Loc	Outdoors	SURF/U-SURF
Dayoub [90]	Omnidir	Topo	Map + Loc	Indoors	SURF
Bacca [91, 92]	Omnidir	Topo	Map + Loc	Indoors	SIFT/SURF
Bacca [93]	Omnidir	Topo	SLAM	Indoors	Edges
Romero [94, 95]	Omnidir	Topo	SLAM	Outdoors	MSER
Majdik [96]	Mono	Topo	Loc	Outdoors	ASIFT
Saedan [97]	Omnidir	Hybrid	SLAM	Indoors	Wavelets
Kessler [98]	Omnidir	Topo	SLAM	Indoors	SIFT
Maohai [99]	Omnidir	Topo	Map + Loc	Indoors	ASIFT

are then made to landmarks instead of individual images, resulting into a dense continuous topological map without sacrificing the speed of the solution. They presented a probabilistic localization approach using the learned discriminative properties of each landmark.

Kawewong et al. presented Position-Invariant Robust Features (PIRFs) [69, 70], a method for generating averaged features from SIFT descriptors that can be matched along several consecutive frames in a temporal window given the input sequence of images. Each place was represented by a dictionary of these representative PIRFs, whose variation of appearance was assumed relatively small with regard to robot motion. These features were then used in an incremental appearance-based SLAM algorithm called PIRF-Nav, which was based on a majority voting scheme. Despite they showed several improvements in terms of recall regarding other common solutions, the main problem of this approach was the computational cost, since some images took long time to be processed. In order to improve this performance, Tongprasit et al. [71] modified the original PIRF algorithm and added a new dictionary management in a SLAM approach called PIRF-Nav 2. This method was 12 times faster than the original PIRF-Nav sacrificing only a small percentage of recall. Morioka et al. [72] presented a method for mapping PIRFs in three-dimensional space combining them with an odometry source. Their method, called 3D-PIRF, was validated navigating in crowded indoor environments.

Andreasson and Duckett [73] presented a simplified version of the SIFT algorithm (M-SIFT) adapted to omnidirectional images, where the descriptors are only found in one resolution, because full invariance to scale and translation is not required in their case. Interest points are selected using the Shi and Tomasi method. Several image description methods used for topological localization were presented, showing the M-SIFT approach the best performance with regard to the other ones. Using the M-SIFT descriptor, Valgren et al. [74] represented the environment by means of an image similarity matrix. They avoided exhaustively computing the affinity matrix by searching for cells which are more likely to describe existing loop closures. Later, in [75], they employed exhaustive search, but introduced an incremental spectral clustering algorithm to reduce the search space incrementally when new images are processed. They also addressed the topological localization problem for outdoor environments over time [76], comparing SIFT and SURF for these purposes and concluding that SURF performs better for topological localization in outdoor scenarios. Moreover, Ascani et al. [77] found that SIFT performs better in indoor environments for topological localization tasks. Other authors that created a topological map from a similarity matrix are Anati and Daniilidis [78]. In their work, they introduced a novel image similarity measure for panoramas which involves dynamic programming to match images using both the appearance and the relative positions of local features simultaneously. The probability of loop closures is modelled using a Markov Random Field (MRF) over the image similarity matrix.

Some researchers construct hierarchical maps of the environment from a set of input images. These approaches combine higher level conceptual maps (usually topological) with lower level and geometrically accurate maps, trying to maximize the advantages and minimize the problems of each kind of map alone and combine them in a different mapping technique. Zivkovic et al. [79] presented an algorithm for automatically generating hierarchical maps from images. A low-level map is built using SIFT features and geometrical constraints. They then use the graph-cuts algorithm to cluster nodes to construct a high-level representation. This hierarchical representation was later employed in [80], where they showed a navigation system based on a topological space which used the epipolar geometry and a planar floor constraint to obtain a heading estimation. This work was further improved in [81] proposing a incremental data association scheme based on the concept of Connected Dominating Set (CDS) of a graph. Given a new image, this method is used to find a subset of past images that represents the complete image set, enabling an efficient loop closure detection during the trajectory of the robot. Dayoub et al. [82] presented a solution where an initial dense pose-graph map of the environment were generated using a graph-based SLAM algorithm. This map is then used to infer a sparse hybrid map with two levels, global and local. The global level is represented by a topological map built using a dual clustering approach. On the local level, each node stores a spherical view representation of the features extracted from images recorded at the position of the node, which is used for estimating the robot's heading using a multiple-view geometry approach. As a contribution of this book, in Chap. 7, a hybrid vision-based topological mapping approach is introduced.

Instead of inferring a high-level topological map from a set of geometric relations, other authors have proposed an alternative hybrid representation where each node of a global topological map includes its own metric sub-map. Blanco et al. [83] presented an approach called Hybrid Metric-Topological SLAM (HTM-SLAM). The sequence of areas traversed by the robot is modelled as a graph whose nodes are annotated with metric sub-maps and whose arcs include the coordinate transformation between these areas. They also proposed a unified Bayesian approach to estimate the robot's path while traversing the environment. This work was improved in [84] using spectral techniques to efficiently partition the map into sub-maps and deriving expressions for applying their ideas to other sensors, such as a stereo camera. In the same line, Tully et al. [85] proposed a hybrid localization solution based on the *hierarchical atlas map* [101], a structure specially created for robots operating in large environments. In this framework, a global topological map decomposes the space into regions within which a feature-based map is built. The localization process is separated in two steps. First, a discrete probability distribution is computed using a recursive Bayesian filter in order to determine the most probable map. Next, a metric position is estimated within the correspondent sub-map using a Kalman filter. Later, in [86], they investigated SLAM as a multi-hypothesis topological loop closing problem. Both works were combined in a more complete solution recently in [102].

Segvic et al. [87] created a hybrid visual navigation framework for large-scale mapping and localization combining several features extracted from monocular perspective images. Despite the approach supported navigation based exclusively on 2D image measurements, it relied in 3D reconstruction procedures. Ramisa et al. [88] also tried to combine several local feature region detectors in order to create a signature of a place for localization purposes. They showed that these combinations increase notably the performance compared with the use of one descriptor alone. Badino et al. [89] integrated metric data directly into a topological map in their hybrid approach called *topometric* localization. Each node of the graph is stored together with its GPS position. They grab images at a constant Euclidean distance, and for each one, visual local features are extracted. A feature database is generated next, where each feature is stored with a reference to the node corresponding to its real location. This database is then used by a Bayes filter to estimate the probability density function of the position of the observer as the vehicle moves along the route.

The multi-store model of human memory proposed by Atkinson and Shiffrin [103] has inspired several approaches. This model divides the human memory into three stores: Sensory Memory (SM), Short-Term Memory (STM) and Long-Term Memory (LTM). Input information is stored in the SM. A selective attention process determines which information can be moved to the STM. Information stored in this memory can be forgotten as soon as it is no longer attended to. Through a rehearsal process, information is moved from the STM to the LTM in order to be retained for longer periods. Dayoub and Duckett [90] used these concepts in order to keep up to date the appearance of a particular place in a map in response to the dynamic changes of the environment during a long-term operation. Bacca et al. [91, 92] adapted this human memory model considering a weighted voting scheme. This allows to pass to the STM only strong features present in the environment. The memory model is implemented using a Feature Stability Histogram (FSH), which stores information about the number of times each feature has been observed in each node. A more complete FSH approach was presented in [93], adapting the initial solution to operate in SLAM conditions.

Romero and Cazorla [94, 95] proposed an approach to construct topological maps matching graphs of invariant features. Each image is segmented into regions in order to group the extracted invariant features in a graph so that each graph defines a single region of the image. The matching process takes into the account the features and their structure using the Graph Transformation Matching (GTM) algorithm.

Recently, Majdik et al. [96] dealt with the *air-ground* matching localization problem, where images taken by a camera mounted on a Micro Aerial Vehicle (MAV) need to be matched with a set of images stored in a database of geotagged pictures obtained from Google Street View. To overcome the severe viewpoint changes presented, they proposed to generate virtual views of each scene, exploiting the air-ground geometry of the system. The best image correspondences are obtained using a histogram-voting scheme. They compared their solution with several state-of-the-art approaches, outperforming them in computational terms and precision-recall rates.

Other solutions based on local features [97, 98] included particle filters as a method to estimate the probability distribution of the location over the topological map. More recently, Maohai et al. [99] combined a particle filter with a GPU-based image description and matching algorithm to define a complete topological autonomous navigation system for indoor environments.

3.4 Methods Based on Bag-of-Words Schemes

The Bag-of-Words (BoW) algorithm has recently been used in a high number of different topological mapping approaches due to its ability for rapidly finding similar image candidates in large datasets of images. The reader is referred to Sect. 2.2.2 for further details about the BoW model.

Visual vocabularies employed in the BoW model are usually generated offline in a training phase. As will be seen in Sect. 3.6, generating the visual dictionary in an offline phase presents several problems. In order to overcome these drawbacks, some authors have proposed to build it in an incremental fashion, adapting the codewords to the appearance of the operating scenario. In this section, the BoW-based works are classified according to this criterion. The main approaches based on BoW schemes are summarized in Table 3.3 following the same guidelines as the previous sections.

3.4.1 Offline Visual Vocabulary Approaches

Despite the BoW algorithm has been used in other areas, such as for internet search engines or for scene categorization [143, 144], it was first applied to visual search techniques in the seminal work of Sivic an Zisserman [145], where this model was employed in order to find similar scenes in video sequences. SIFT features were extracted from each frame and then quantized as BoW vectors, creating a database of BoW image representations. They presented an interactive application where the user could query the image database to find similar frames, i.e. with enough features in common. A lookup table called inverted file, which mapped image words to the video frames where they were found, was also used to speed up the retrieval process. Wang et al. [104, 105] presented a coarse-to-fine global localization system based on the BoW model, where interest points detected with the Harris-Laplace detector were described using the SIFT algorithm. In an offline phase, the vocabulary and the inverted index were created, and then used for localization. An epipolar geometry step was incorporated in order to verify whether the loop candidate obtained from the BoW stage was plausible.

The size of a dictionary can vary within a large range, which has an impact on the performance of the retrieval process. The larger the size, the more discriminating the vocabulary is, but at a higher computational cost for finding the nearest reference descriptor. The hierarchical visual vocabulary has been proposed as a relevant improvement towards alleviating this problem [146], where the original training set of descriptors is clustered in a small number of clusters, and then each cluster is recursively clustered again until achieving the desired number of words. Given a query descriptor, finding its closest word consists in traversing the tree from the root until reaching a leaf node. This hierarchical representation, in addition to the inverted index, makes the BoW algorithm an ideal and scalable approach for searching

Table 3.3 Summary of topological mapping and localization solutions based on BoW schemes

References	Camera	Map	Tasks	Environment	Features
Wang [104, 105]	Mono	Topo	Map + Loc	In + Out	HARRIS/SIFT
Fraundorfer [106]	Mono	Topo	Map + Loc	Indoors	MSER/SIFT
Konolige [107]	Stereo	Hybrid	SLAM	In + Out	STAR/FAST/SAD
Cummins [108, 109]	Mono	Topo	SLAM	Outdoors	SIFT/SURF
Cummins [110, 111]	Mono	Topo	SLAM	Outdoors	SURF
Cummins [112, 113]	Omnidir	Topo	SLAM	Outdoors	SURF
Newman [114]	Omnidir	Hybrid	SLAM	Outdoors	SURF
Maddern [115, 116]	Omnidir	Hybrid	SLAM	Outdoors	SURF
Maddern [117]	Omnidir	Hybrid	SLAM	Indoors	SURF
Paul [118]	Mono	Topo	SLAM	Outdoors	SURF
Johns [119, 120]	Mono	Topo	SLAM	Outdoors	SIFT
Galvez [121, 122]	Mono	Topo	SLAM	In + Out	FAST/BRIEF
Mur-Artal [123]	Mono	Topo	SLAM	Outdoors	ORB
Ranganathan [124]	Mono	Hybrid	SLAM	Indoors	SIFT
Cadena [125]	Stereo	Topo	SLAM	In + Out	SURF
Ciarfuglia [126]	Mono	Topo	SLAM	In + Out	SURF
Majdik [127]	Mon/Ste	Topo	SLAM	Outdoors	SURF
Schindler [128]	Mono	Topo	Map + Loc	Outdoors	SIFT
Achar [129]	Mono	Topo	Map + Loc	Outdoors	SIFT
Lee [130]	Mono	Topo	SLAM	Indoors	MSLD
Filliat [131]	Mono	Topo	Map + Loc	Indoors	SIFT
Angeli [132]	Mono	Topo	SLAM	Indoors	SIFT
Angeli [133]	Mono	Topo	SLAM	In + Out	SIFT/Color Hist.
Angeli [134]	Mono	Topo	SLAM	Indoors	SIFT/Color Hist.
Labbe [135, 136]	Mono	Topo	SLAM	In + Out	SURF
Nicosevici [137, 138]	Mono	Topo	SLAM	Underwater	SURF
Khan [139]	Mono	Topo	SLAM	In + Out	BRISK
Murphy [140]	Mono	Topo	SLAM	In + Out	-
MacTavish [141]	Mono	Hybrid	Map + Loc	Outdoors	SURF
Mohan [142]	Mono	Hybrid	Map + Loc	In + Out	ORB

millions of images in an efficient way and it is a good option to consider when mapping large environments. Fraundorfer et al. [106] applied this hierarchical dictionary to the visual navigation problem, presenting a highly scalable vision-based localization and mapping method using image collections. For each frame captured by the camera, they used the dictionary structure and the inverted file to retrieve the most likely images. Using a RANSAC procedure, they performed a geometry verification step against these candidates, which can be used to determine if the image closes a loop or otherwise is a new place to be added to the map. They used the local geometric information to navigate within the generated topological map. Konolige et al. [107] proposed a SLAM solution based on an adapted scheme of this hierarchical codebook using a stereo camera. As shown in their results, the approach, which was assessed in indoor and outdoors environments, was able to find loop closures in paths of several kilometres. A strong geometric filter was used to eliminate false positives when detecting loop closures.

Probably the most well-known solution that falls into this category is the Cummins and Newman's Fast Appearance-Based Mapping (FAB-MAP) approach [108, 109], proposed under the assumption that modelling the probabilities that the visual words appear simultaneously can help in the localization process. These probabilities were approximated by a Chow Liu tree, computed from a set of training data as the maximum-weight spanning tree of a directed graph of co-occurrences between visual words. This approximation permitted the authors to compute efficiently an observation likelihood which was used in a Bayes filter for predicting loop closure candidates. The main drawback presented by the original FAB-MAP algorithm was the high computational cost, since every time the robot collected an observation, the likelihood needed to be computed for each location existent in the map. To solve this problem, Cummins and Newman [110, 111] introduced a probabilistic bail-out test based on the use of concentration inequalities for rapidly identifying promising loop closure hypotheses and then avoid to compute the likelihood for all locations. Later, an even faster version called FAB-MAP 2.0 [112, 113] was presented adapting the probabilistic model to be used with an inverted index architecture similar to image typical search engines. This scheme was assessed using a dataset of 1,000 km composed by omnidirectional images and GPS coordinates to be used as ground truth. FAB-MAP was combined with a laser in the work of Newman et al. [114], where it was used as a component to detect loop closures in urban scenes.

Initially, the authors only published FAB-MAP as binaries to the community. For this reason, Glover et al. developed OpenFABMAP [147], a fully open-source implementation of the algorithm, adding some improvements. OpenFABMAP was a key component in the solution proposed by Maddern et al. called Continuous Appearance-based Trajectory SLAM (CAT-SLAM) [115, 116], where an appearance-based SLAM system was improved with odometric information using a particle filter in order to obtain an estimation of the position of the vehicle. An extension of CAT-SLAM called CAT-Graph was introduced in [117] combining multiple visits to the same place to build a topological graph-based representation of indoor environments. These graphs were used in the mapping and localization processes according to the loop closures detected by the appearance-based module.

Since the BoW model used in FAB-MAP does not take into account the spatial arrangement of the visual words, Paul and Newman introduced FAB-MAP 3D [118], where they demonstrated that integrating this kind of information in the algorithm improved the localization accuracy. Using a random graph, they modelled the word co-occurrences as well as their pairwise distances and showed how to accelerate the inference process with a Delaunay tessellation of this graph. Another attempt to include spatial information within the BoW model for localization is the recent work by Johns and Yang, where they presented the Feature Co-occurrence Maps (Cooc-Map) [119], where local features are quantized in both feature and image space and a set of statistics regarding their co-occurrence at different times of the day are calculated. They also introduce a new geometric feature matching algorithm for this kind of representation and showed how sequential matching can be incorporated into their solution. They also showed that learning the properties of local features observed during long periods of time can be more accurate for localization than representing a location using a single image [120].

An attempt to create a visual dictionary from binary features can be found in the work of Galvez-Lopez and Tardos [121, 122]. They adapted the hierarchical BoW model of Nister to be used with keypoints detected with FAST and described with the BRIEF algorithm. Other novelties of their work included a direct index to obtain correspondences between images in a efficient manner and matching images in groups to increase the accuracy of the loop closure detection process. Using this framework, they are able to detect loop closures in sequences of 19,000 images spending an average time of 16 ms per image, presenting an interesting improvement in performance in comparison to other solutions. Their dictionary-building approach was recently used in combination with the ORB descriptor in [123], showing improvements in the recognition performance.

Ranganathan et al. presented an approach called Online Probabilistic Topological Mapping (OPTM) [124], an online loop-closing algorithm based on a Rao-Blackwellized particle filter which was used for updating incrementally the posterior on the space of all possible topologies whenever a new measurement arrived. Since OPTM was sensor independent, it was assessed with a laser range finder, an odometry source and visual input in indoor environments. A BoW model based on a multivariate Polya distribution was used for quantizing SIFT descriptors. OPTM improves a previous framework called Probabilistic Topological Maps (PTM) [148] by enhancing the inference process so that it can be used online.

Cadena et al. [125] introduced a place recognition framework based on stereo vision which combined a BoW model for obtaining loop closure candidates and an algorithm based on Conditional Random Fields (CRF-Matching) in order to verify these candidates. This matching method, according to the authors, was more robust than using only epipolar geometry, since it used 3D information provided by the stereo images. This module was later used in [149], where a method for removing past incorrect loop closures using the Realizing, Reversing, Recovering (RRR) algorithm was presented.

Some authors have proposed weighing strategies different to the one typically used in BoW approaches, i.e. TF-IDF. For a start, Ciarfuglia et al. [126] showed a discriminative criterion to assign weights to the visual words in a training phase. The weights are learnt in an approach based on the large margin paradigm and can be applied to several similarity functions in order to compare images. This weighing scheme was assessed in a loop closure detection module within a SLAM framework for navigating in indoor and outdoor environments. Another case is Majdik et al. [127] who proposed an adaptive loop closure algorithm based on the hierarchical BoW model that was able to update the weights of the visual words according to their importance when detecting loop closures. They assessed their approach using both single and stereo cameras in outdoor environments.

While in outdoor environments GPS can be used for estimating the location of a robot, urban environments present more challenging situations since buildings can block the satellite signals. Clearly, vision becomes an option as exteroceptive sensor in these cases. Nevertheless, indexing images from a city can be very difficult in computational terms, reason why the BoW model can be of help for this kind of situations. In line with this scenario, Schindler et al. [128] presented a localization system for recognizing scenes in cities, where they were able to index 30,000 images from a city using a BoW scheme. They showed that this huge amount of information can be more efficiently retrieved by selecting the most informative features from the training dataset, understanding these features as the ones that occur in all images of some specific location but not in other places. This concept was measured using the information gain formula. They also proposed an alternative search algorithm called Greedy N-Best Paths (GNP) improving the image retrieval performance. A more recent solution for urban localization can be found in the work by Achar et al. [129], where geometric inferencing was used to identify features corresponding to moving objects in the scene. These features are then used for global localization.

Recently, Lee et al. [130] proposed a place recognition system that, instead of quantizing interest points, they processed lines using Mean Standard-Deviation Line descriptors (MSLD). A hierarchical visual dictionary was trained using these vectors, which was employed in combination with a Bayes filter for detecting loop closures in indoor environments. They integrated this loop closure detection module into a SLAM solution.

Other place recognition solutions which have appeared recently are based on the BoW framework, adapting FAB-MAP to work as a hierarchical approach [141] or maintaining an inverted file per group of images or environments [142].

3.4.2 Online Visual Vocabulary Approaches

An alternative to maintain the dictionary adapted to the operating environment is to generate it online, at the same time that the robot explores the world. In this regard, Filliat [131] introduced an approach to construct dynamically a visual dictionary. The closest visual word to a given local feature was selected performing a simple linear search algorithm. If these features were very far in distance, the query local feature was added as a new word to the dictionary. This scheme was assessed using different feature spaces and employed for mapping and localization tasks, but it was limited to small distances due to the inefficiency of the linear search algorithm. This model was extended by Angeli et al. [132] to incremental conditions to be used in a place recognition module. Their approach relied on a discrete Bayes filter to estimate the probability of loop closures and to ensure temporal coherency between predictions. During the calculation of the likelihood, the TF-IDF coefficients were extracted according to the distinctiveness of each word given the current image. This work was improved in [133], where two visual vocabularies were trained and used together as input to the Bayes filter, and further expanded in [134] by constructing a complete topological SLAM system.

Inspired by the work of Angeli, Labbe and Michaud presented a solution called Real-Time Appearance-Based Mapping (RTAB-Map) [135, 136] a loop closure detection approach for large-scale and long-term SLAM. The main contribution of this solution was that they provided memory management mechanisms for caching a subset of the online learnt visual words in the main memory (called Working Memory), and this subset was used for detecting loop closures. The rest were stored in a database stored in an external memory called Long Term Memory. The transition of words between memories was ruled by the time taken for processing images in an adaptive way. This scheme allowed to obtain high recall rates at 100% of precision while maintaining the real time performance of the solution.

Nicosevici and Garcia [137, 138] introduced Online Visual Vocabulary (OVV), where the words were generated at the same time that the robot was exploring the environment using a modified version of an agglomerative clustering algorithm. The elementary clusters were created from features that can be tracked along the images of the sequence, represented by the mean descriptor of a feature and the covariance matrix of the observed descriptors at the current point. In order to merge these clusters, they provided a novel criterion based on the Fisher's linear discriminant that took into account the global distribution of the data, resulting into more distinctive visual words. A method for efficiently reindexing the images when the vocabulary changes is also proposed. An interesting aspect of their experimental results is that, in addition to outdoor scenarios, the approach was assessed in underwater environments. The OVV technique was recently used in [140] for performing unsupervised topological place recognition in an image stream captured by a robot.

More recently, an incremental BoW scheme based on binary descriptors called IBuILD [139] has emerged. In this work, the authors propose a method to construct a visual dictionary that can be used for loop closure detection. However, the authors do not use an indexing scheme for an efficient search of features.

Despite they are more related to the pose-graph SLAM field, there exists other solutions that used a BoW scheme built in an online manner that can be interesting for the reader, such as the works of Eade and Drummond [150], Botterill et al. [151] and Pradeep et al. [152].

3.5   Methods Based on Combined Approaches

In order to maximize the benefits of each approach, several authors have proposed solutions based on combinations of different image descriptors for topological mapping and localization. The main approaches that fall into this category are summarized in Table 3.4 specifying the same characteristics as in previous sections.

Table 3.4 Summary of topological mapping and localization solutions based on combined approaches

References	Camera	Map	Tasks	Environment	Combination
Goedeme [153, 154]	Omnidir	Topo	Map + Loc	Indoors	SIFT/Columns
Murillo [155, 156]	Omnidir	Hybrid	Map + Loc	In + Out	SURF/Color Hist.
Wang [157]	Mono	Topo	Map + Loc	In + Out	OACH/SIFT
Weiss [158, 159]	Mono	Topo	Map + Loc	Outdoors	WGOH/WGII/SIFT
Siagian [160]	Mono	Topo	Map + Loc	Outdoors	Gist/SIFT
Chapoulie [161]	Sphere	Topo	SLAM	Outdoors	SIFT/Spatial Hists.
Wang [162]	Omnidir	Topo	Map + Loc	Indoors	SURF/Convex Hull
Lin [163]	Omnidir	Topo	Map + Loc	In + Out	SURF/Convex Hull
Wang [164]	Mono	Topo	Map + Loc	Outdoors	Harris/Color Hist.
Maohai [165]	Omnidir	Hybrid	Map + Loc	Outdoors	Color Hist./SIFT
Korrapati [166, 167]	Omnidir	Topo	Mapping	Outdoors	SURF/BoW

A common approach is to use a global descriptor to perform a fast selection of similar images during an image search and then use a more accurate process in order to confirm the association, such as matching local features. Goedeme et al. [153] presented a localization system for omnidirectional cameras where, for each acquired image, they extracted vertical column segments and described them with ten different descriptors. After a clustering process, these local descriptors were inserted into a kd-tree structure that was used by the localization process. When a query image arrived, the same local descriptors applied to the vertical structures were computed over the entire image and used to rapidly retrieve possible loop candidates. Next, a matching distance based on the column segments was applied between the image and each of the candidates in order to ensure a correct image matching. The localization process was supported by a Bayes filter, which allowed them to deal with noisy measurements. Their work was improved in [154], presenting a complete navigation system, adding SIFT features to the framework and applying the Dempster–Shafer probabilistic theory to the topological map construction.

Murillo et al. [155] proposed a three-step hierarchical localization method for omnidirectional images. A global colour descriptor was applied to obtain a set of susceptible loop candidates, and then line features described by their line support regions were matched using pyramidal matching in order to find the most similar image given a predefined visual memory. The 1D radial trifocal tensor was employed to obtain a metric localization. Their work was expanded incorporating SURF features to the framework [156].

Wang and Yagi [157] combined recently their OACH global descriptor with local features extracted with the Harris-Laplace detector and described by the SIFT descriptor. They created two databases: one for OACH descriptors for coarse localization and a SIFT database for fine localization. During the global localization stage, a set of candidate images was extracted and then a fine localization step against this subset was performed. A RANSAC-based fundamental matrix estimation strategy was employed in order to verify if the image association was correct.

Weiss et al. [158] performed outdoor localization using a particle filter where particle weights were updated according to the similarities computed using two global descriptors: WGOH and WGII. To calculate the similarity between two images, each descriptor was compared independently using normalized histogram intersection and the final distance was the product of the previous results. This method was compared with SIFT, presenting a slightly minor recall, but four times faster. Later in [159], SIFT was incorporated into their framework as an alternative to compute the position of the robot in those cases where it can not be inferred using the combined global descriptors method.

Another localization approach based on particle filters and inspired in biological concepts can be found in the work proposed by Siagian and Itti [160], which is based in Gist and saliency features, implemented in parallel using shared raw feature channels.

Chapoulie et al. [161] introduced a loop closing algorithm to be used with spherical images. SIFT features were extracted as local features, while histograms of their distribution over the features space were used as global features. These representations were combined in a Bayes filter in order to detect loop closure candidates under outdoor environments.

Wang and Lin presented a combined local and global descriptor for omnidirectional images called Hull Census Transform (HCT) [162], which consisted of repeatedly generating the convex hull from the extracted SURF features and computing the relative magnitude between these features that compose the convex hull, resulting into a set of binary vectors. This representation was then used for detecting scene changes, generating a set of topological node lists. This work was recently expanded by Lin et al. [163] in a new combined descriptor called Extended-HTC, where they included colour information from the environment, encoded as colour histograms, as well as the structure information of the convex hulls, computed by means of the centroid of the features and the total distance between any two feature point locations.

A location recognition system which combined edges, local features and colour histograms was proposed by Wang and Yagi [164]. The image description process was computed in an integrated way: the Harris detector was used to obtain both edges and interests points, while SIFT algorithm was used for describing interest points.

Maohai et al. [165] presented a hierarchical localization approach based on omnidirectional vision where, in a first step, colour histograms allow to select a subset of the images stored in the map. Next, SIFT local features are used to obtain a more accurate localization inside this subset.

Recently, Korrapati et al. [166] presented a hierarchical mapping model which organized images into a topological map using the Vector of Locally Aggregated Descriptors (VLAD), where the quantization residues of the local features descriptors, such as SURF, were combined into a single descriptor. This allowed them to create maps containing over 11,000 images and a decent amount of frames per second. In a more recent work [167], they also proposed a hierarchical topological mapping algorithm using a sparse node representation where Hierarchical Inverted Files (HIF) were employed for an efficient two-level map storage.

3.6 Discussion

In the last decades, there has been a significant increase in the number of visual solutions for topological mapping and localization because of the low cost of cameras and the richness of the sensor data provided. This chapter surveyed the main approaches emerged in the last fifteen years. We identified that these works can be classified, according to the method used for representing the image, into four main categories:

- methods based on global descriptors, where the image is represented by a general descriptor computed using the entire visual information as input;
- methods based on local descriptors, where interest points are found in the image and then a patch around this point is described in order to identify them in other images;
- methods based on the BoW algorithm, where local features are quantized according to a set of feature models called visual dictionary, representing images as histograms of occurrences of each word in the image; and
- methods based on combined descriptors, where several techniques described above are used together as a new solution.

The main advantages and disadvantages of each method are summarized in Table 3.5. All these methods are active research areas and authors publish continuously solutions for mapping, localization or SLAM facing the problem from the point of view of these approaches.

Regarding the different categories of methods enumerated above, global descriptors are normally very fast to compute, favouring the matching process between the images and reducing the computational needs of mapping and localization tasks. As a main disadvantages, they offer less robustness to occlusion and illumination effects, what results in a lower discriminative power and an increment of the perceptual aliasing effect, where different places can be perceived as the same. They have been used intensively in other related research areas, such as scene categorization.

Table 3.5 Advantages and disadvantages of each method. More '∗' means better performance regarding the corresponding attribute

Feature	Global descriptors	Local features	BoW schemes
CPU needs	∗ ∗ ∗	∗	∗∗
Storage needs	∗ ∗ ∗	∗	∗∗
Matching complexity	∗∗	∗	∗ ∗ ∗
Discrimination power	∗	∗ ∗ ∗	∗∗
Perceptual aliasing effect	∗	∗ ∗ ∗	∗∗
Large-scale operation	∗∗	∗	∗ ∗ ∗
Spatial loss information	∗∗	∗ ∗ ∗	∗
Pose recovery complexity	∗	∗ ∗ ∗	∗∗

Local features are usually more robust to occlusions and changes in scale, rotation and illumination. These methods start with a detection phase, where interest points are found in the image, and are followed by a description phase, where some measures are extracted from the surroundings of these keypoints. Local features present a better discrimination capacity, resulting into higher recognition rates and less detection errors. Furthermore, the recovery of relative poses between images, which can be used for confirming if two images come from the same scene, can be performed easily. However, the storage requirements and the computational cost are higher than for global descriptors and the matching process is also more complex, since sometimes each query descriptor requires to find their closest neighbour within a large set of features. According to the surveyed works, the most used feature is SIFT, followed by SURF, both representing features as vectors of floating point numbers. Recently, a number of binary descriptors have been proposed in the literature, providing an interesting research line to explore regarding topological mapping and localization, because they are cheaper to compute, compact to store and faster to compare.

While global descriptors and local features demonstrate useful approaches for robot mapping and localization, they do not result to be satisfactory when the number of images to process is high. Matching hundreds of images using local features can take a long time when trying to associate the current frame with every previously seen location. Indexing structures can be used to accelerate the search. However, with a high number of descriptors, memory problems and computational bottlenecks appear. Global descriptors are easier to compute and save storage space, but sacrificing discriminative power which reduces the performance of the solution. In this case, an alternative approach for describing and matching images is the Bag-Of-Words (BoW) algorithm, which can efficiently index a huge amount of images incorporating a hierarchical scheme and an inverted index structure. Due to this fast image retrieval, works classified in this category are mainly SLAM approaches. As main limitation, it can be mentioned the fact that the effect of perceptual aliasing worsens due to the quantization process, the presence of noisy words due to the coarseness of the vocabulary construction method and the loss of the spatial relations between the words. Some authors have proposed several improvements in order to overcome this last drawback [119, 168].

The visual dictionaries can be generated offline or online. As a main shortcoming, the offline approaches need a training phase, where sometimes millions of descriptors have to be clustered. This can take hours, depending on the number of images and the clustering technique used. Furthermore, the robot can operate in an environment with an appearance totally different to the training set employed for generating the dictionary, which implies that it is not representative of the scenario, increasing false detections. An alternative is to build the codebook online in an incremental manner, while the robot is navigating across the environment. However, this implies inserting and deleting features to/from the dictionary, limiting its possible size. An interesting study about the reuse of visual dictionaries and their universality is presented by Hou et al. [169]. Nowadays, despite several approaches have been proposed, managing efficiently online visual dictionaries for BoW schemes can be considered as a topic of interest. Another interesting issue is long-term mapping, in order to manage maps

Table 3.6 Advantages and disadvantages of methods for generating visual dictionaries in BoW schemes

Feature	Offline	Online
Training phase	Needed	No
Scenario model updated	No	Yes
Incremental memory management	No	Needed
Dictionary size	Large	As required
Handle large dictionaries	Yes	As required

during long periods of time under changes in the appearance of the environment. The main advantages and limitations of each dictionary-generation approach are summarized in Table 3.6.

After the deep literature review performed in this chapter and considering the open research topics found, in this book, three visual topological mapping approaches are proposed, tackling the problem from different points of view:

- In Chap. 5, the problem of indexing floating-point local features efficiently for topological mapping is addressed, resulting into a solution called FEATMap.
- In Chap. 6, we propose an incremental Bag-of-Binary-Words scheme for place recognition called OBIndex, and, next, this approach is used as a key component in a dense topological mapping solution called BINMap.
- In Chap. 7, to further improve the results obtained with BINMap and to favour long-term mapping tasks, a hierarchical approach called HTMap is introduced.

The approaches presented in this monograph are evaluated using a common framework, which is presented in the following chapter.

References

1. Bonin-Font, F., Ortiz, A., Oliver, G.: Visual navigation for mobile robots: a survey. J. Intell. Robot. Syst. **53**(3), 263–296 (2008)
2. Fuentes-Pacheco, J., Ruiz-Ascencio, J., Rendón-Mancha, J.M.: Visual simultaneous localization and mapping: a survey. Artif. Intell. Rev. **43**(1), 55–81 (2015)
3. Olson, E., Leonard, J., Teller, S.: Fast iterative alignment of pose graphs with poor initial estimates. IEEE Int. Conf. Robot. Autom. 2262–2269 (2006)
4. Frese, U., Schroder, L.: Closing a million-landmarks loop. IEEE/RSJ Int. Conf. Intell. Robot. Syst. 5032–5039 (2006)
5. Dellaert, F., Kaess, M.: Square root sam: simultaneous localization and mapping via square root information smoothing. Int. J. Robot. Res. **25**(12), 1181–1203 (2006)
6. Kaess, M., Ranganathan, A., Dellaert, F.: iSAM: incremental smoothing and mapping. IEEE Trans. Robot. **24**(6), 1365–1378 (2008)

7. Grisetti, G., Stachniss, C., Burgard, W.: Nonlinear constraint network optimization for efficient map learning. IEEE Trans. Intell. Transp. Syst. **10**(3), 428–439 (2009)

8. Konolige, K., Grisetti, G., Kummerle, R., Burgard, W., Limketkai, B., Vincent, R.: Efficient sparse pose adjustment for 2d mapping. IEEE/RSJ Int. Conf. Intell. Robot. Syst. 22–29 (2010)

9. Kaess, M., Johannsson, H., Roberts, R., Ila, V., Leonard, J., Dellaert, F.: iSAM2: incremental smoothing and mapping with fluid relinearization and incremental variable reordering. IEEE Int. Conf. Robot. Autom. 3281–3288 (2011)

10. Kummerle, R., Grisetti, G., Strasdat, H., Konolige, K., Burgard, W.: g2o: A general framework for graph optimization. IEEE Int. Conf. Robot. Autom. 3607–3613 (2011)

11. Wu, J., Christensen, H., Rehg, J.: Visual place categorization: problem, dataset, and algorithm. IEEE/RSJ Int. Conf. Intell. Robot. Syst. 4763–4770 (2009)

12. Winters, N., Gaspar, J., Lacey, G., Santos-Victor, J.: Omni-directional vision for robot navigation. In: IEEE workshop on omnidirectional vision, pp. 21–28 (2000)

13. Gaspar, J., Winters, N., Santos-Victor, J.: Vision-based navigation and environmental representations with an omnidirectional camera. IEEE Trans. Robot. Autom. **16**(6), 890–898 (2000)

14. Ulrich, I., Nourbakhsh, I.: Appearance-based place recognition for topological localization. IEEE Int. Conf. Robot. Autom. **2**, 1023–1029 (2000)

15. Werner, F., Maire, F., Sitte, J.: Topological slam using fast vision techniques. Advances in Robotics, pp. 187–196. Springer, Berlin (2009)

16. Kosecka, J., Zhou, L., Barber, P., Duric, Z.: Qualitative image based localization in indoors environments. IEEE Conf. Comput. Vis. Pattern Recog. **2**, II-3–II-8 (2003)

17. Bradley, D., Patel, R., Vandapel, N., Thayer, S.: Real-time image-based topological localization in large outdoor environments. IEEE/RSJ Int. Conf. Intell. Robot. Syst. 3670–3677 (2005)

18. Weiss, C., Masselli, A.: Fast outdoor robot localization using integral invariants. IEEE Int. Conf. Comput. Vis. 1–10 (2007)

19. Wang, J., Zha, H., Cipolla, R.: Efficient topological localization using orientation adjacency coherence histograms. Int. Conf. Pattern Recog. 271–274 (2006)

20. Pronobis, A., Caputo, B., Jensfelt, P., Christensen, H.: A discriminative approach to robust visual place recognition. IEEE/RSJ Int. Conf. Intell. Robot. Syst. 3829–3836 (2006)

21. Singh, G., Kosecka, J.: Visual loop closing using gist descriptors in manhattan world. In: IEEE Workshop on Omnidirectional Vision, Camera Networks and Non-classical Camera (2010)

22. Murillo, A.C., Campos, P., Kosecka, J., Guerrero, J.: Gist vocabularies in omnidirectional images for appearance based mapping and localization. In: Workshop on Omnidirectional Vision, Camera Networks and Non-classical Cameras (RSS) (2010)

23. Rituerto, A., Murillo, A.C., Guerrero, J.: Semantic labeling for indoor topological mapping using a wearable catadioptric system. Robot. Auton. Syst. **62**, 685–695 (2013)

24. Sunderhauf, N., Protzel, P.: BRIEF-gist - closing the loop by simple means. IEEE/RSJ Int. Conf. Intell. Robot. Syst. 1234–1241 (2011)

25. Arroyo, R., Alcantarilla, P.F., Bergasa, L.M., Yebes, J., Gamez, S.: Bidirectional loop closure detection on panoramas for visual navigation. Intell. Vehic. Symp. 1378–1383 (2014)

26. Arroyo, R., Alcantarilla, P.F., Bergasa, L.M., Yebes, J.J., Bronte, S.: fast and effective visual place recognition using binary codes and disparity information. IEEE/RSJ Int. Conf. Intell. Robot. Syst. (2014)

27. Liu, Y., Zhang, H.: Visual loop closure detection with a compact image descriptor. IEEE/RSJ Int. Conf. Intell. Robot. Syst. 1051–1056 (2012)

28. Chapoulie, A., Rives, P., Filliat, D.: Topological segmentation of indoors/outdoors sequences of spherical views. IEEE/RSJ Int. Conf. Intell. Robot. Syst. 4288–4295 (2012)

29. Chapoulie, A., Rives, P., Filliat, D.: Appearance-based segmentation of indoors and outdoors sequences of spherical views. IEEE Int. Conf. Robot. Autom. 1946–1951 (2013)

30. Lamon, P., Nourbakhsh, I., Jensen, B., Siegwart, R.: Deriving and matching image fingerprint sequences for mobile robot localization. IEEE Int. Conf. Robot. Autom. **2**, 1609–1614 (2001)

31. Tapus, A., Tomatis, N., Siegwart, R.: Topological global localization and mapping with fin-gerprints and uncertainty. In: International symposium on experimental robotics, pp. 18–21 (2004)
32. Tapus, A., Siegwart, R.: Incremental robot mapping with fingerprints of places. IEEE/RSJ Int. Conf. Intell. Robot. Syst. 2429–2434 (2005)
33. Liu, M., Scaramuzza, D., Pradalier, C., Siegwart, R., Chen, Q.: Scene recognition with omni-directional vision for topological map using lightweight adaptive descriptors. IEEE/RSJ Int. Conf. Intell. Robot. Syst. 116–121 (2009)
34. Liu, M., Siegwart, R.: DP-FACT: Towards topological mapping and scene recognition with color for omnidirectional camera. IEEE Int. Conf. Robot. Autom. 3503–3508 (2012)
35. Menegatti, E., Maeda, T., Ishiguro, H.: Image-based memory for robot navigation using properties of omnidirectional images. Robot. Auton. Syst. **47**(4), 251–267 (2004)
36. Menegatti, E., Zoccarato, M., Pagello, E., Ishiguro, H.: Image-based monte carlo localisation with omnidirectional images. Robot. Auton. Syst. **48**(1), 17–30 (2004)
37. Payá, L., Fernández, L., Gil, A., Reinoso, O.: Map building and monte carlo localization using global appearance of omnidirectional images. Sensors **10**(12), 11468–11497 (2010)
38. Ranganathan, A., Menegatti, E., Dellaert, F.: Bayesian inference in the space of topological maps. IEEE Trans. Robot. **22**(1), 92–107 (2006)
39. Milford, M., Wyeth, G., Prasser, D.: RatSLAM: A hippocampal model for simultaneous localization and mapping. In: IEEE Int. Conf. Robot. Autom. 403–408 (2004)
40. Prasser, D., Milford, M., Wyeth, G.: Outdoor simultaneous localisation and mapping using RatSLAM. FSR 143–154 (2005)
41. Milford, M., Wyeth, G.: Mapping a suburb with a single camera using a biologically inspired SLAM system. IEEE Trans. Robot. **24**(5), 1038–1053 (2008)
42. Glover, A., Maddern, W., Milford, M., Wyeth, G.: FAB-MAP + RatSLAM: appearance-based slam for multiple times of day. IEEE Int. Conf. Robot. Autom. 3507–3512 (2010)
43. Lui, W.L.D., Jarvis, R.: A pure vision-based approach to topological SLAM. IEEE/RSJ Int. Conf. Intell. Robots Syst. 784–3791 (2010)
44. Lui, W.L.D., Jarvis, R.: A pure vision-based topological SLAM system. Int. J. Robot. Res. **31**(4), 403–428 (2012)
45. Badino, H., Huber, D., Kanade, T.: Real-time topometric localization. IEEE Int. Conf. Robot. Autom. 1635–1642 (2012)
46. Xu, D., Badino, H., Huber, D.: Topometric localization on a road network. IEEE/RSJ Int. Conf. Intell. Robot. Syst. (2014)
47. Lategahn, H., Beck, J., Kitt, B., Stiller, C.: How to learn an illumination robust image feature for place recognition. Intell. Vehic. Symp. 285–291 (2013)
48. Nourani-Vatani, N., Borges, P., Roberts, J., Srinivasan, M.: On the use of optical flow for scene change detection and description. J. Intell. Robot. Syst. **74**(3), 817–846 (2014)
49. Milford, M., Wyeth, G.: SeqSLAM: Visual route-based navigation for sunny summer days and stormy winter nights. IEEE Int. Conf. Robot. Autom. 1643–1649 (2012)
50. Milford, M.: Visual route recognition with a handful of bits. Robot. Sci. Syst. (2013)
51. Milford, M.: Vision-based place recognition: how low can you go? Int. J. Robot. Res. **32**(7), 766–789 (2013)
52. Pepperell, E., Corke, P., Milford, M.: All-environment visual place recognition with SMART. IEEE Int. Conf. Robot. Autom. 1612–1618 (2014)
53. Wu, J., Zhang, H., Guan, Y.: An efficient visual loop closure detection method in a map of 20 million key locations. IEEE Int. Conf. Robot. Autom. 861–866 (2014)
54. Oliva, A., Torralba, A.: Modeling the shape of the scene: a holistic representation of the spatial envelope. Int. J. Comput. Vis. **42**(3), 145–175 (2001)
55. Siagian, C., Itti, L.: Rapid biologically-inspired scene classification using features shared with visual attention. IEEE Trans. Pattern Anal. Mach. Intell. **29**(2), 300–12 (2007)
56. Calonder, M., Lepetit, V., Strecha, C., Fua, P.: BRIEF: binary robust independent elementary features. European Conference on Computer Vision. Lecture Notes in Computer Science, vol. 6314, pp. 778–792. Springer, Berlin (2010)

57. Yang, X., Cheng, K.T.: Local difference binary for ultrafast and distinctive feature description. IEEE Trans. Pattern Anal. Mach. Intell. **36**(1), 188–94 (2014)
58. Prasser, D., Wyeth, G.: Probabilistic visual recognition of artificial landmarks for simultaneous localization and mapping. IEEE Int. Conf. Robot. Autom. **1**, 1291–1296 (2003)
59. Agrawal, M., Konolige, K., Blas, M.R.: CenSurE: center surround extremas for realtime feature detection and matching. European Conference on Computer Vision, vol. 5305, pp. 102–115. Springer, Berlin (2008)
60. Kosecka, J., Yang, X.: Location recognition and global localization based on scale-invariant keypoints. In: Workshop on Statistical Learning in Computer Vision (ECCV) (2004)
61. Kosecka, J., Li, F.: Vision based topological markov localization. IEEE Int. Conf. Robot. Autom. **2**, 1481–1486 (2004)
62. Li, F., Kosecka, J.: Probabilistic location recognition using reduced feature set. IEEE Int. Conf. Robot. Autom. 405–3410 (2006)
63. Zhang, H.: BoRF: loop-closure detection with scale invariant visual features. IEEE Int. Conf. Robot. Autom. 3125–3130 (2011)
64. Zhang, H.: Indexing visual features: real-time loop closure detection using a tree structure. IEEE Int. Conf. Robot. Autom. 3613–3618 (2012)
65. Rybski, P., Zacharias, F., Lett, J.F., Masoud, O., Gini, M., Papanikolopoulos, N.: Using visual features to build topological maps of indoor environments. IEEE Int. Conf. Robot. Autom. **1**, 850–855 (2003)
66. He, X., Zemel, R., Mnih, V.: Topological map learning from outdoor image sequences. J. Field Robot. **23**(11–12), 1091–1104 (2006)
67. Sabatta, D.G.: vision-based topological map building and localisation using persistent features. In: Robotics and mechatronics symposium, pp. 1–6 (2008)
68. Johns, E., Yang, G.Z.: Global localization in a dense continuous topological map. IEEE Int. Conf. Robot. Autom. 1032–1037 (2011)
69. Kawewong, A., Tangruamsub, S., Hasegawa, O.: Position-invariant robust features for long-term recognition of dynamic outdoor scenes. IEICET. Inf. Syst. **E93-D**(9), 2587–2601 (2010)
70. Kawewong, A., Tongprasit, N., Tangruamsub, S., Hasegawa, O.: Online and incremental appearance-based SLAM in highly dynamic environments. Int. J. Robot. Res. **30**(1), 33–55 (2011)
71. Tongprasit, N., Kawewong, A., Hasegawa, O.: PIRF-Nav 2: speeded-up online and incremental appearance-based slam in an indoor environment. In: IEEE Workshop on Applications of Computer Vision, pp. 145–152 (2011)
72. Morioka, H., Yi, S., Hasegawa, O.: Vision-based mobile robot's slam and navigation in crowded environments. IEEE/RSJ Int. Conf. Intell. Robot. Syst. 3998–4005 (2011)
73. Andreasson, H., Duckett, T.: Topological localization for mobile robots using omnidirectional vision and local features. IFAC Symp. Intell. Auton. Vehic. (2008)
74. Valgren, C., Lilienthal, A., Duckett, T.: Incremental topological mapping using omnidirectional vision. IEEE/RSJ Int. Conf. Intell. Robot. Syst. 3441–3447 (2006)
75. Valgren, C., Duckett, T., Lilienthal, A.: Incremental spectral clustering and its application to topological mapping. IEEE Int. Conf. Robot. Autom. 10–14 (2007)
76. Valgren, C., Lilienthal, A.: SIFT, SURF and seasons: long-term outdoor localization using local features. Eur. Conf. Mob. Rob. **128**, 1–6 (2007)
77. Ascani, A., Frontoni, E., Mancini, A., Zingaretti, P.: Feature group matching for appearance-based localization. IEEE/RSJ Int. Conf. Intell. Robot. Syst. 3933–3938 (2008)
78. Anati, R., Daniilidis, K.: constructing topological maps using markov random fields and loop-closure detection. In: Advances in Neural Information Processing Systems, pp. 37–45 (2009)
79. Zivkovic, Z., Bakker, B., Krose, B.: Hierarchical map building using visual landmarks and geometric constraints. IEEE/RSJ Int. Conf. Intell. Robot. Syst. 2480–2485 (2005)
80. Booij, O., Terwijn, B., Zivkovic, Z., Krose, B.: Navigation using an appearance based topological map. IEEE Int. Conf. Robot. Autom. 3927–3932 (2007)

81. Booij, O., Zivkovic, Z., Krose, B.: Efficient data association for view based slam using connected dominating sets. Robot. Auton. Syst. **57**(12), 1225–1234 (2009)
82. Dayoub, F., Cielniak, G., Duckett, T.: A sparse hybrid map for vision-guided mobile robots. In: Eur. Conf. Mob. Robot. 213–218 (2011)
83. Blanco, J.L., Fernandez-Madrigal, J.A., Gonzalez, J.: Towards a unified bayesian approach to hybrid metric-topological SLAM. IEEE Trans. Robot. **24**(2), 259–270 (2008)
84. Blanco, J.L., Gonzalez, J., Fernandez-Madrigal, J.A.: Subjective local maps for hybrid metric-topological SLAM. Robot. Auton. Syst. **57**(1), 64–74 (2009)
85. Tully, S., Moon, H., Morales, D., Kantor, G., Choset, H.: Hybrid localization using the hierarchical atlas. IEEE/RSJ Int. Conf. Intell. Robot. Syst. 2857–2864 (2007)
86. Tully, S., Kantor, G., Choset, H., Werner, F.: A multi-hypothesis topological slam approach for loop closing on edge-ordered graphs. IEEE/RSJ Int. Conf. Intell. Robot. Syst. 4943–4948 (2009)
87. Segvic, S., Remazeilles, A., Diosi, A., Chaumette, F.: A mapping and localization framework for scalable appearance-based navigation. Comput. Vis. Image Und. **113**(2), 172–187 (2009)
88. Ramisa, A., Tapus, A., Aldavert, D., Toledo, R.: Lopez de Mantaras: R.: Robust vision-based robot localization using combinations of local feature region detectors. Auton. Robot. **27**(4), 373–385 (2009)
89. Badino, H., Huber, D., Kanade, T.: Visual topometric localization. Intell. Vehic. Symp. 794–799 (2011)
90. Dayoub, F., Duckett, T.: an adaptive appearance-based map for long-term topological localization of mobile robots. IEEE/RSJ Int. Conf. Intell. Robot. Syst. 3364 – 3369 (2008)
91. Bacca, B., Salvi, J., Batlle, J., Cufi, X.: Appearance-based mapping and localisation using feature stability histograms. Elect. Lett. **46**(16), 1120 (2010)
92. Bacca, B., Salvi, J., Cufi, X.: Appearance-based mapping and localization for mobile robots using a feature stability histogram. Robot. Auton. Syst. **59**(10), 840–857 (2011)
93. Bacca, B., Salvi, J., Cufi, X.: Long-term mapping and localization using feature stability histograms. Robot. Auton. Syst. **61**(12), 1539–1558 (2013)
94. Romero, A., Cazorla, M.: Topological SLAM using omnidirectional images: merging feature detectors and graph-matching. Advanced Concepts for Intelligent Vision Systems. Lecture Notes in Computer Science, vol. 6474, pp. 464–475. Springer, Berlin (2010)
95. Romero, A., Cazorla, M.: Topological visual mapping in robotics. Cogn. Process. **13**(1), 305–308 (2012)
96. Majdik, A., Albers-Schoenberg, Y., Scaramuzza, D.: MAV urban localization from google street view data. IEEE/RSJ Int. Conf. Intell. Robot. Syst. 3979–3986 (2013)
97. Saedan, M., Lim, C.W., Ang, M.: Appearance-based slam with map loop closing using an omnidirectional camera. IEEE Int. Conf. Adv. Intell. Mech. 1–6 (2007)
98. Kessler, J., König, A., Gross, H.M.: An improved sensor model on appearance based SLAM. Auton. Mob. Syst. **216487**, 153–160 (2009)
99. Maohai, L., Han, W., Lining, S., Zesu, C.: Robust omnidirectional mobile robot topological navigation system using omnidirectional vision. Eng. Appl. Artif. Intell. **26**(8), 1942–1952 (2013)
100. Zhang, H., Li, B., Yang, D.: Keyframe detection for appearance-based visual SLAM. IEEE/RSJ Int. Conf. Intell. Robot. Syst. 2071–2076 (2010)
101. Lisien, B., Morales, D., Silver, D., Kantor, G., Rekleitis, I.M., Choset, H.: The hierarchical atlas. IEEE Trans. Robot. **21**(3), 473–481 (2005)
102. Tully, S., Kantor, G., Choset, H.: A unified bayesian framework for global localization and SLAM in hybrid metric/topological maps. Int. J. Robot. Res. **31**(3), 271–288 (2012)
103. Atkinson, R.C., Shiffrin, R.M.: Human memory: a proposed system and its control processes. Psychol. Learn. Motiv. Adv. Res. Theory **2**, 89–105 (1968)
104. Wang, J., Cipolla, R., Zha, H.: Vision-based global localization using a visual vocabulary. IEEE Int. Conf. Robot. Autom. 4230–4235 (2005)
105. Wang, J., Zha, H., Cipolla, R.: Coarse-to-fine vision-based localization by indexing scale-invariant features. IEEE Trans. Syst. Man Cybern. Part B Cybern. **36**(2), 413–422 (2006)

106. Fraundorfer, F., Engels, C., Nister, D.: Topological mapping, localization and navigation using image collections. IEEE/RSJ Int. Conf. Intell. Robot. Syst. 3872–3877 (2007)
107. Konolige, K., Bowman, J., Chen, J., Mihelich, P., Calonder, M., Lepetit, V., Fua, P.: View-based maps. Int. J. Robot. Res. **29**(8), 941–957 (2010)
108. Cummins, M., Newman, P.: Probabilistic appearance based navigation and loop closing. IEEE Int. Conf. Robot. Autom. 2042–2048 (2007)
109. Cummins, M., Newman, P.: FAB-MAP: probabilistic localization and mapping in the space of appearance. Int. J. Robot. Res. **27**(6), 647–665 (2008)
110. Cummins, M., Newman, P.: Accelerated appearance-only SLAM. IEEE Int. Conf. Robot. Autom. 1828–1833 (2008)
111. Cummins, M., Newman, P.: Accelerating FAB-MAP with concentration inequalities. IEEE Trans. Robot. **26**(6), 1042–1050 (2010)
112. Cummins, M., Newman, P.: Highly scalable appearance-only SLAM - FAB-MAP 2.0. Robot. Sci. Syst. 1–8 (2009)
113. Cummins, M., Newman, P.: Appearance-only SLAM at large scale with FAB-MAP 2.0. Int. J. Robot. Res. **30**(9), 1100–1123 (2011)
114. Newman, P., Sibley, G., Smith, M., Cummins, M., Harrison, A., Mei, C., Posner, I., Shade, R., Schroeter, D., Murphy, L., Churchill, W., Cole, D., Reid, I.: Navigating, recognizing and describing urban spaces with vision and lasers. Int. J. Robot. Res. **28**(11–12), 1406–1433 (2009)
115. Maddern, W., Milford, M., Wyeth, G.: Continuous appearance-based trajectory SLAM. IEEE Int. Conf. Robot. Autom. 3595–3600 (2011)
116. Maddern, W., Milford, M., Wyeth, G.: CAT-SLAM: probabilistic localisation and mapping using a continuous appearance-based trajectory. Int. J. Robot. Res. **31**(4), 429–451 (2012)
117. Maddern, W., Milford, M., Wyeth, G.: Towards persistent indoor appearance-based localization, mapping and navigation using CAT-graph. IEEE/RSJ Int. Conf. Intell. Robot. Syst. 4224–4230 (2012)
118. Paul, R., Newman, P.: FAB-MAP 3D: Topological mapping with spatial and visual appearance. IEEE Int. Conf. Robot. Autom. 2649–2656 (2010)
119. Johns, E., Yang, G.Z.: Feature co-occurrence maps: appearance-based localisation throughout the day. IEEE Int. Conf. Robot. Autom. 3212–3218 (2013)
120. Johns, E., Yang, G.Z.: Dynamic scene models for incremental, long term, appearance based localisation. IEEE Int. Conf. Robot. Autom. 2731–2736 (2013)
121. Galvez-Lopez, D., Tardos, J.: Real-time loop detection with bags of binary words. IEEE/RSJ Int. Conf. Intell. Robot. Syst. 51–58 (2011)
122. Galvez-Lopez, D., Tardos, J.: Bags of binary words for fast place recognition in image sequences. IEEE Trans. Robot. **28**(5), 1188–1197 (2012)
123. Mur-Artal, R., Tardos, J.D.: Fast relocalisation and loop closing in keyframe-based SLAM. IEEE Int. Conf. Robot. Autom. 846–853 (2014)
124. Ranganathan, A., Dellaert, F.: Online probabilistic topological mapping. Int. J. Robot. Res. **30**(6), 755–771 (2011)
125. Cadena, C., Galvez-Lopez, D., Ramos, F., Tardos, J., Neira, J.: Robust place recognition with stereo cameras. IEEE/RSJ Int. Conf. Intell. Robot. Syst. 5182–5189 (2010)
126. Ciarfuglia, T., Costante, G., Valigi, P., Ricci, E.: A discriminative approach for appearance based loop closing. IEEE/RSJ Int. Conf. Intell. Robot. Syst. 3837–3843 (2012)
127. Majdik, A., Galvez-Lopez, D., Lazea, G., Castellanos, J.: Adaptive appearance based loop-closing in heterogeneous environments. IEEE/RSJ Int. Conf. Intell. Robot. Syst. 1256–1263 (2011)
128. Schindler, G., Brown, M., Szeliski, R.: City-scale location recognition. IEEE Conf. Comput. Vis. Pattern Recog. 1–7 (2007)
129. Achar, S., Jawahar, C., Madhava Krishna, K.: Large scale visual localization in urban environments. IEEE Int. Conf. Robot. Autom. 5642–5648 (2011)
130. Lee, J.H., Zhang, G., Lim, J., Suh, I.H.: Place recognition using straight lines for vision-based SLAM. IEEE Int. Conf. Robot. Autom. 3799–3806 (2013)

131. Filliat, D.: A visual bag of words method for interactive qualitative localization and mapping. IEEE Int. Conf. Robot. Autom. 3921–3926 (2007)
132. Angeli, A., Doncieux, S., Meyer, J.A., Filliat, D.: Real-time visual loop-closure detection. IEEE Int. Conf. Robot. Autom. 1842–1847 (2008)
133. Angeli, A., Filliat, D., Doncieux, S., Meyer, J.A.: A fast and incremental method for loop-closure detection using bags of visual words. IEEE Trans. Robot. **24**(5), 1027–1037 (2008)
134. Angeli, A., Doncieux, S., Meyer, J.A., Filliat, D.: Incremental vision-based topological SLAM. IEEE/RSJ Int. Conf. Intell. Robot. Syst. 1031–1036 (2008)
135. Labbe, M., Michaud, F.: Memory management for real-time appearance-based loop closure detection. IEEE/RSJ Int. Conf. Intell. Robot. Syst. 1271–1276 (2011)
136. Labbe, M., Michaud, F.: Appearance-based loop closure detection for online large-scale and long-term operation. IEEE Trans. Robot. **29**(3), 734–745 (2013)
137. Nicosevici, T., Garcia, R.: On-line visual vocabularies for robot navigation and mapping. IEEE/RSJ Int. Conf. Intell. Robot. Syst. 205–212 (2009)
138. Nicosevici, T., Garcia, R.: Automatic visual bag-of-words for online robot navigation and mapping. IEEE Trans. Robot. **28**(4), 886–898 (2012)
139. Khan, S., Wollherr, D.: IBuILD: Incremental bag of binary words for appearance-based loop closure detection. IEEE Int. Conf. Robot. Autom. 5441–5447 (2015)
140. Murphy, L., Sibley, G.: Incremental unsupervised topological place discovery. IEEE Int. Conf. Robot. Autom. 1312–1318 (2014)
141. MacTavish, K., Barfoot, T.D.: Towards hierarchical place recognition for long-term autonomy. IEEE Int. Conf. Robot. Autom. (2014)
142. Mohan, M., Galvez-Lopez, D., Monteleoni, C., Sibley, G.: Environment selection and hierarchical place recognition. IEEE Int. Conf. Robot. Autom. (2015)
143. Csurka, G., Dance, C., Fan, L., Willamowski, J., Bray, C.: Visual categorization with bags of keypoints. Eur. Conf. Comput. Vis. **1**, 1–22 (2004)
144. Li, F.F., Perona, P.: A bayesian hierarchical model for learning natural scene categories. IEEE Conf. Comput. Vis. Pattern Recog. 524–531 (2005)
145. Sivic, J., Zisserman, A.: Video google: a text retrieval approach to object matching in videos. IEEE Int. Conf. Comput. Vis. 1470–1477 (2003)
146. Nister, D., Stewenius, H.: Scalable recognition with a vocabulary tree. IEEE Conf. Comput. Vis. Pattern Recog. **2**, 2161–2168 (2006)
147. Glover, A., Maddern, W., Warren, M., Reid, S., Milford, M., Wyeth, G.: OpenFABMAP: an open source toolbox for appearance-based loop closure detection. IEEE Int. Conf. Robot. Autom. 4730 – 4735 (2012)
148. Ranganathan, A., Dellaert, F.: a rao-blackwellized particle filter for topological mapping. IEEE Int. Conf. Robot. Autom. 810–817 (2006)
149. Latif, Y., Cadena, C., Neira, J.: Realizing, reversing, recovering: incremental robust loop closing over time using the irrr algorithm. IEEE/RSJ Int. Conf. Intell. Robot. Syst. 4211–4217 (2012)
150. Eade, E., Drummond, T.: Unified loop closing and recovery for real time monocular SLAM. British Mach. Vis. Conf. 1–10 (2008)
151. Botterill, T., Mills, S., Green, R.: Bag-of-words-driven, single-camera simultaneous localization and mapping. J. Field Robot. **28**(2), 204–226 (2011)
152. Pradeep, V., Medioni, G., Weiland, J.: Visual loop closing using multi-resolution SIFT Grids in metric-topological SLAM. IEEE Conf. Comput. Vis. Pattern Recog. 1438–1445 (2009)
153. Goedemé, T., Nuttin, M., Tuytelaars, T., Van Gool, L.: Markerless computer vision based localization using automatically generated topological maps. In: European Navigation Conference, pp. 235–243 (2004)
154. Goedemé, T., Nuttin, M., Tuytelaars, T., Van Gool, L.: Omnidirectional vision based topological navigation. Int. J. Comput. Vis. **74**(3), 219–236 (2007)
155. Murillo, A.C., Sagues, C., Guerrero, J.: From omnidirectional images to hierarchical localization. Robot. Auton. Syst. **55**(5), 372–382 (2007)

156. Murillo, A.C., Guerrero, J., Sagues, C.: SURF features for efficient robot localization with omnidirectional images. IEEE Int. Conf. Robot. Autom. 3901–3907 (2007)
157. Wang, J., Yagi, Y.: Efficient Topological Localization Using Global and Local Feature Matching. Int. J. Adv. Robot. Syst. **10**(153:2013), – (2013)
158. Weiss, C., Masselli, A., Zell, A.: fast vision-based localization for outdoor robots using a combination of global image features. IFAC Symp. on Intell. Auton. Vehic. 119–124 (2007)
159. Weiss, C., Tamimi, H., Masselli, A., Zell, A.: A hybrid approach for vision-based outdoor robot localization using global and local image features. IEEE/RSJ Int. Conf. Intell. Robot. Syst. 1047–1052 (2007)
160. Siagian, C., Itti, L.: Biologically inspired mobile robot vision localization. IEEE Trans. Robot. **25**(4), 861–873 (2009)
161. Chapoulie, A., Rives, P., Filliat, D.: A spherical representation for efficient visual loop closing. IEEE Int. Conf. Comput. Vis. 335–342 (2011)
162. Wang, M.L., Lin, H.Y.: A hull census transform for scene change detection and recognition towards topological map building. IEEE/RSJ Int. Conf. Intell. Robot. Syst. 548–553 (2010)
163. Lin, H.Y., Lin, Y.H., Yao, J.W.: Scene Change Detection and Topological Map Construction Using Omnidirectional Image Sequences. In: IAPR Int. Conf. Mach. Vision App. 56–60 (2013)
164. Wang, J., Yagi, Y.: Robust location recognition based on efficient feature integration. IEEE Int. Conf. Robot. Biomim. 97–101 (2012)
165. Maohai, L., Lining, S., Qingcheng, H., Zesu, C., Songhao, P.: Robust omnidirectional vision based mobile robot hierarchical localization and autonomous navigation. Inf. Tech. J. **10**(1), 29–39 (2011)
166. Korrapati, H., Uzer, F., Mezouar, Y.: Hierarchical visual mapping with omnidirectional images. IEEE/RSJ Int. Conf. Intell. Robot. Syst. 3684–3690 (2013)
167. Korrapati, H., Mezouar, Y.: Vision-based sparse topological mapping. Robot. Auton. Syst. **62**(9), 1259–1270 (2014)
168. Lazebnik, S., Schmid, C., Ponce, J.: Beyond bags of features: spatial pyramid matching for recognizing natural scene categories. IEEE Conf. Comput. Vis. Pattern Recog. **2**, 2169–2178 (2006)
169. Hou, J., Liu, W.X., E, X., Xia, Q., Qi, N.M.: An experimental study on the universality of visual vocabularies. J. Vis. Commun. Image. Represent. **24**(7), 1204–1211 (2013)

Chapter 4
Experimental Setup

Abstract The topological mapping algorithms presented in this book have been validated using a common framework, featuring several criteria for performance evaluation and a number of relevant public datasets representing different scenarios of operation. The algorithms have as well been compared against some state-of-the-art solutions. The goal of this chapter is to summarize this experimental framework, which will be used to evaluate the solutions proposed in the rest of the book.

4.1 Performance Metrics

Loop closure detection can be seen as a binary classification problem, where the classifier is the algorithm itself and its output represents if the current image closes a loop with an already seen image or not. Therefore, it is usual to validate the performance of a loop closure algorithm employing methods which are normally used to evaluate binary classifiers.

Binary classification is the task of distributing the elements of a set in two groups according to a discrimination rule. This kind of classification has been applied successfully in different tasks, such as medical testing or information retrieval. According to the classification rules, a classifier can produce errors, which can be stated as:

- false positive errors (FP), commonly referred to as *false alarms*, which are produced when the result of the classification indicates the presence of a condition whereas actually it is not fulfilled, and,
- false negative errors (FN), which are produced when the result of the classification indicates that a condition is not fulfilled, whereas actually it is.

Conversely, a correct result can be classified into true positives (TP) or true negatives (TN) depending on the presence of the condition or not. In order to associate the response of a classifier to one of these groups, we need information provided by direct observation in contrast to the result itself, which has been inferred. This information is commonly referred to as the *ground truth*.

E. Garcia-Fidalgo and A. Ortiz, *Methods for Appearance-based Loop Closure Detection*, Springer Tracts in Advanced Robotics 122, https://doi.org/10.1007/978-3-319-75993-7_4

It is interesting to note that in many practical binary classification problems, the two groups are not symmetric and then the relative proportion between the different types of errors is of interest. For instance, in medical testing, false positives (detecting a disease when it is not present) and false negatives (not detecting a disease when it is present) are considered different error cases. Then, it is common to take this fact into account to evaluate the performance of a binary classifier, obtaining ratios between the different types of errors instead of total numbers.

There exist several metrics and ratios to evaluate the performance of a binary classifier, and the election depends on the field of application [1]. In computer science and information retrieval, the combination of *precision* and *recall* is usually the preferred option. Precision is the fraction of positive retrieved instances that are actually positive, while recall is the fraction of positive retrieved instances out of the total number of positive instances:

$$P = \frac{TP}{TP + FP}, \tag{4.1}$$

$$R = \frac{TP}{TP + FN}. \tag{4.2}$$

Informally, a high precision value means a low number of false positives, while a high recall value means a low number of false negatives. A graphical representation of the precision and recall metrics is shown in Fig. 4.1.

In order to use the precision-recall metrics to evaluate the performance of the algorithms presented in this book, the datasets introduced in Sect. 4.2 are provided with a ground truth, which indicates, for each image in the sequence, which other images can be considered to close a loop with it. Then, the assessment is performed counting for each dataset the number of true positives, true negatives, false positives and false negatives, where positive is meant for detection of loop closure. The precision P is then defined as the ratio of real loop closure detections to total amount of loop closures detected, while the recall R is defined as the ratio of real loop closures to total amount of loop closures existent in the dataset.

For loop closure detection, it is essential to avoid false positives, since it means that two images have been identified as a loop but, in reality, they represent different places. This will induce the algorithm to produce inconsistent maps and, therefore, avoiding these false positives becomes essential. By definition, if no false positives are found, the precision reaches 100% (see Eq. 4.1). Then, in our experiments, we are interested in finding the best recall than can be achieved at 100% of precision using each approach, which indicates the percentage of loop closures that can be detected by the algorithm without false positive detections.

Another common performance measure used in this work is the precision-recall (PR) curve. A PR curve is a graphical plot that illustrates the performance of the algorithm as a critical parameter is varied. As the name suggests, the curve is created by plotting the precision (P) against the recall (R) obtained for different values of that parameter. This plot is useful to validate the sensitivity and the behaviour

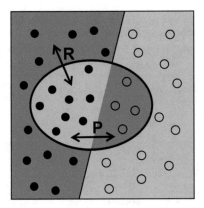

Fig. 4.1 Precision and recall metrics. The items that are positive according to the ground truth are located to the left of the straight line, while the items retrieved as positive by the binary classifier are inside the oval area. The red areas represent errors. Then, the red area located to the left of the line and outside of the oval area represents the positive items that could not be retrieved (false negatives), while the red area inside the oval area represents the items retrieved as positives that are not actually positives (false positives)

of the algorithms with regard to a critical parameter with a great impact in the global performance of the approach, such as an acceptance threshold. PR curves are usually presented as an alternative to Receiver Operating Characteristic (ROC) curves depending on the class distribution [2]. Unlike ROC curves, the best performance in a PR curve is indicated with 100% of precision and 100% of recall (the upper-right corner of the plot), which also yields to the maximum Area Under the Curve (AUC).

The execution times for each component of the different algorithms are also measured, obtaining the average, the standard deviation, the maximum and the minimum values. This allows us to validate the performance of the solutions according to their computational times, giving an estimation of the time needed to execute each of them. All experiments were performed on a laptop fitted with an Intel Core i3 at 2.27 GHz processor and 8 GB of RAM memory and running GNU/Linux Ubuntu. Regarding the implementation of the different solutions, all algorithms presented in this book were developed using the C++ programming language, the OpenCV library for image-related operations and the OpenMP library to execute processes in parallel, specially in Chap. 8. The corresponding source codes were reasonably optimized.

4.2 Datasets

In order to carry out the performance evaluation of the solutions proposed in this book, several datasets have been employed. A total of 10 image sequences from 4 different datasets were selected, including indoor and outdoor environments. These

Table 4.1 Sequence of images used in this book. The total distance travelled in the indoor sequences is unknown, since GPS signal is not available in these scenarios

Sequence	#Images	Img size (px)	Rate (Hz)	Dist (km)	Env
Lip6 Indoor [3]	388	240×192	1.0	Unknown	Indoors
Lip6 Outdoor [3]	1063	240×192	1.0	1.41	Outdoors
City centre [4]	1237	1280×480	0.5	2.01	Outdoors
New college [4]	1073	1280×480	0.5	1.92	Outdoors
KITTI 00 [5]	4541	1241×376	10.0	3.73	Outdoors
KITTI 05 [5]	2761	1226×370	10.0	2.22	Outdoors
KITTI 06 [5]	1101	1226×370	10.0	1.27	Outdoors
UIBSmallLoop [6]	388	300×240	0.5	0.47	Outdoors
UIBLargeLoop [6]	997	300×240	0.5	1.48	Outdoors
UIBIndoor [6]	384	300×240	0.5	Unknown	Indoors

sequences, which are summarized in Table 4.1, have been used independently and in no particular order in the following chapters. In this section, the main characteristics of each dataset are enumerated.

4.2.1 Lip6 Dataset

Lip6 is a public dataset[1] recorded for validating the work of Angeli et al. [3]. It consists of two sequences with images of size 240 × 192: one indoors and one outdoors. The indoor sequence, referred to as *Lip6 Indoor*, performs two loops inside a building under high perceptual aliasing conditions while the outdoor sequence, referred to as *Lip6 Outdoor*, presents a very changing scenario around a building at the city of Paris. Both sequences were recorded with a hand-held camera approximately pointing in the direction of displacement and were acquired at 1 Hz. The authors provide ground truth files for each sequence, which were created by hand by themselves. Some examples of images from the Lip6 sequences are shown in Fig. 4.2.

4.2.2 Oxford Dataset

The Oxford dataset[2] consists of two different sequences: *City Centre* and *New College*. These sequences were originally obtained for the evaluation of FAB-MAP [4] and consist of, respectively, 1237 and 1073 pairs of images of size 640 × 480 taken by

---

[1]http://cogrob.ensta-paristech.fr/loopclosure.html.

[2]http://www.robots.ox.ac.uk/~mobile/IJRR_2008_Dataset.

Fig. 4.2 Examples of images taken from the Lip6 Indoor sequence (top row) and the Lip6 Outdoor sequence (bottom row)

two cameras mounted on a robot while it travels through the city of Oxford. Since the approaches presented in this book have been developed to be used with monocular cameras, we merge left and right frames resulting into images of size 1280×480. The City Centre sequence was recorded to validate the ability of a system for matching images in the presence of scene changes, while the New College sequence was recorded because of its high perceptual aliasing conditions. Along with the images, the authors provide ground truth files labelled by their own observations. Unlike the Lip6 dataset, the positions of the images are also available. This allows us to plot the detected loop closures and the resulting maps when using this dataset, as will be seen in the following chapters. Some examples of images taken from these sequences are shown in Fig. 4.3.

4.2.3   KITTI Dataset

The third group of sequences have been obtained from the Karlsruhe Institute of Technology and Toyota Technological Institute (KITTI[3]) suite [5]. This suite comprises benchmarks for stereo, optical flow, visual odometry, SLAM and 3D object detection and was recorded using a platform equipped with four high resolution video cameras, a Velodyne laser scanner and a localization system. In this book, we use the odometry benchmark. More precisely, the KITTI odometry benchmark consists of 22 outdoor sequences, with more than 40000 images covering a total of 39.2 km. Among these 22 sequences, there are 12 that contain loop closures. We employ sequences

---

[3]http://www.cvlibs.net/datasets/kitti.

Fig. 4.3 Examples of images taken from the City Center (top) and the New College (bottom) sequences

Fig. 4.4 Examples of images taken from the KITTI 00 (top), KITTI 05 (middle) and KITTI 06 sequences (bottom)

00, 05 and 06 as a representative set of this benchmark. Since originally the KITTI sequences did not include a specific ground truth for loop closure detection, we use the ones provided by Arroyo et al. [7]. Each dataset is also provided with a pose file, which is exclusively used to plot the positions of the images. Examples of these sequences are shown in Fig. 4.4.

4.2.4 UIB Dataset

The UIB dataset has been recorded by ourselves during the development of this work. It comprises three sequences: *UIBSmallLoop*, *UIBLargeLoop* and *UIBIndoor*. The UIBSmallLoop and UIBLargeLoop datasets were recorded around the Anselm Turmeda building at the University of the Balearic Islands campus. They consist of 388 and 997 images, respectively, taken under bad weather conditions, for verifying the performance of the approaches under these situations. The UIBIndoor dataset, recorded inside the Anselm Turmeda building, comprises a total of 384 images from an indoor environment which means a number of challenges for loop closure. First of all, the camera velocity is not constant. This is due to the fact that we needed to climb up and down the stairs during the recording. This difficulty enables us to validate the capability of the filter to self-adapt under camera speed changes. Besides, a number of images of white walls result when the camera is at the stairs, what gives rise to the detection of very few features. Moreover, the dataset presents some parts with substantial illumination changes, what leads on some occasions to overexposed images. The ground truth files were created manually by ourselves. These sequences do not include pose files, and then the image positions could not be plotted when using this dataset. Some example images of these sequences are shown in Fig. 4.5.

Fig. 4.5 Examples of images taken from the UIBIndoor sequence (top row) and UIB outdoor sequences (bottom row)

4.3   Reference Solutions

In Chap. 3, we reviewed the most important techniques presented during the last fifteen years in vision-based topological mapping and localization. In this book, two of them are used as a baseline for validating the proposed solutions: FAB-MAP 2.0 [8] and SeqSLAM [9].

4.3.1   FAB-MAP 2.0

FAB-MAP is the best known example of an appearance-based solution implemented using an offline Bag-of-Words (BoW) scheme. FAB-MAP 2.0 [8] is an evolution of the original FAB-MAP algorithm [4], proposed under the assumption that modelling the probabilities of the co-occurrences of visual words can help during the localization process. These probabilities are approximated by a Chow-Liu tree, computed from a set of training data as the maximum-weight spanning tree of a directed graph of co-occurrences between visual words. This approximation allows the authors to compute efficiently a likelihood which is then used in a Bayes filter. FAB-MAP 2.0 includes some enhancements such as the introduction of an inverted file, like in image retrieval systems.

In order to obtain precision-recall metrics from FAB-MAP 2.0, we execute the binaries provided by the authors. The algorithm is configured with the default parameters and using the indoor vocabulary for Lip6 Indoor and UIBIndoor sequences, and the outdoor vocabulary for the rest of sequences. The output is a matrix, the n-th row of which corresponds to the probability distribution over previously seen places due to the n-th image. In this matrix, the main diagonal corresponds to the probability that the image comes from a new place. Since we do not take into account this case in any of our approaches and we want to avoid the false detection of loop closures with recent frames, the output matrix is rectified by removing the most recent probabilities for each row and normalizing the final distribution. A loop closure is detected if the probability is above a predefined threshold p. The precision-recall curves are obtained modifying this threshold p.

4.3.2   SeqSLAM

The second reference work taken into account in this book for comparison purposes is the Milford and Wyeth's SeqSLAM algorithm [9]. This approach is one of the most popular solutions based on global descriptors. SeqSLAM, instead of searching for a single previously seen image given the current frame, performs the localization process recognizing coherent sequences of local consecutive images, even under weather or season changes. They employed normalized patches in a cropped version of the original image, and Sum of Absolute Differences (SAD) to compare these patches.

For testing SeqSLAM, we use the source code provided by OpenSeqSLAM [10]. OpenSeqSLAM has been also configured with the default parameters, except the temporal length of the image sequences (d_s) which is, according to the authors, the most influential parameter of the algorithm. Longer sequence lengths usually perform better in terms of precision-recall, but, in some datasets, they can result into the opposite behaviour because of the absence of sequences of that length, specially in environments with frequent turns. Since we want to increment the performance of each approach, this parameter was experimentally set to 30, what maximized the recall in all datasets. The precision-recall curves are obtained modifying the acceptance threshold.

References

1. Ortiz, A., Oliver, G.: On the use of the overlapping area matrix for image segmentation evaluation: a survey and new performance measures. Pattern Recognit. Lett. **27**(16), 1916–1926 (2006)
2. Davis, J., Goadrich, M.: The relationship between precision-recall and ROC curves. In: ICMLA, pp. 233–240 (2006)
3. Angeli, A., Filliat, D., Doncieux, S., Meyer, J.A.: A fast and incremental method for loop-closure detection using bags of visual words. IEEE Trans. Robot. **24**(5), 1027–1037 (2008)
4. Cummins, M., Newman, P.: FAB-MAP: probabilistic localization and mapping in the space of appearance. Int. J. Robot. Res. **27**(6), 647–665 (2008)
5. Geiger, A., Lenz, P., Urtasun, R.: Are we ready for autonomous driving? The KITTI vision benchmark suite. In: IEEE Conference on Computer Vision Pattern Recognition, pp. 3354–3361 (2012)
6. Garcia-Fidalgo, E., Ortiz, A.: Vision-based topological mapping and localization by means of local invariant features and map refinement. Robotica **33**(7), 1446–1470 (2015)
7. Arroyo, R., Alcantarilla, P.F., Bergasa, L.M., Yebes, J.J., Bronte, S.: Fast and effective visual place recognition using binary codes and disparity information. In: IEEE/RSJ International Conference on Intelligent Robots and Systems (2014)
8. Cummins, M., Newman, P.: Appearance-only SLAM at large scale with FAB-MAP 2.0. Int. J. Robot. Res. **30**(9), 1100–1123 (2011)
9. Milford, M., Wyeth, G.: SeqSLAM: visual route-based navigation for sunny summer days and stormy winter nights. In: IEEE International Conference on Robotics and Automation, pp. 1643–1649 (2012)
10. Sünderhauf, N., Neubert, P., Protzel, P.: Are we there yet? challenging SeqSLAM on a 3000 km journey across all four seasons. In: Workshop on Long-Term Autonomy, ICRA (2013)

Chapter 5
Loop Closure Detection Using Local Invariant Features and Randomized KD-Trees

Abstract This chapter introduces an appearance-based approach for topological mapping and localization named *FEATMap* (Feature-based Mapping). FEATMap relies on a loop closure detection scheme which makes use of local invariant features to describe images. These features are indexed using a set of randomized kd-trees, which permit seeking for matchings between the current and previous images to detect loop closures in a straightforward way. A discrete Bayes filter is added to the solution to obtain loop candidates while ensuring the temporal coherence between consecutive predictions. Finally, FEATMap comprises a method for refining the resulting maps as they are obtained, removing spurious nodes in accordance to the visual information that they contain.

5.1 Overview

As stated in Chap. 3, the Bag-of-Words (BoW) approach [1] is one of the most used techniques in appearance-based loop closure detection. However, this method presents some drawbacks. On the one hand, the quantization process performed during the clustering step contributes to emphasize the perceptual aliasing effect [2], i.e. two different places are perceived as the same because of the similarity between their representations. On the other hand, the training phase is typically performed offline and can take a long time, depending on the number of training descriptors. Global descriptors could be an alternative but usually they are not descriptive enough to be used in an accurate loop detection process.

Topological maps obtained from visual information tend to contain spurious paths and nodes [3, 4]. This is because of image noise, partial invariance to viewpoint, scale and/or illumination changes of image descriptors, or due to the mapping algorithm itself. The final map obtained can be very large and can contain more nodes than are actually required to represent the environment, resulting in an increment of the storage needs and the computational requirements.

To cope with the aforementioned issues, this chapter discusses an appearance-based approach for topological mapping and localization named *FEATMap* (Feature-based Mapping). FEATMap is based on a loop closure detection framework which

© Springer International Publishing AG 2018   69
E. Garcia-Fidalgo and A. Ortiz, *Methods for Appearance-based Loop Closure Detection*, Springer Tracts in Advanced Robotics 122,
https://doi.org/10.1007/978-3-319-75993-7_5

uses local invariant features directly as image description method. To optimize running times, matchings between the current image and previously visited places are determined using an index of features based on a set of randomized kd-trees (see Sect. 2.2.2.1), which is a simple structure that allows us to index images as they are processed, avoiding the classical training step of offline BoW schemes. We use a discrete Bayes filter for predicting loop closure candidates, whose observation model is a novel approach based on an efficient matching scheme between features. In order to avoid redundant information in the resulting maps, we also present a map refinement strategy, which takes into account the visual information stored in the map for refining the resulting final topology. These refined maps save storage space and improve the execution times of localizations tasks. FEATMap is validated under different scenarios and compared with the state-of-the-art FAB-MAP 2.0 algorithm.

The chapter is organized as follows: Sect. 5.2 introduces the image description technique used in FEATMap, Sect. 5.3 describes the structure of the map employed in FEATMap, Sect. 5.4 describes our topological mapping framework in detail, Sect. 5.5 reports on the results of the different experiments performed, and Sect. 5.6 concludes the chapter.

5.2 Image Description

FEATMap uses a kd-tree-based algorithm for indexing features. These algorithms assume the features exist in a real vector space where each dimension of the features can be continuously averaged. For this reason, binary descriptors like BRIEF [5], BRISK [6], ORB [7], FREAK [8] or LDB [9] cannot be used in FEATMap, despite their lower detection and description times. This kind of descriptors will be intensively used in the following chapters, due to their demonstrated benefits. Instead, FEATMap can only employ real-valued descriptors, such as Scale-Invariant Feature Transform (SIFT) [10] or Speeded Up Robust Features (SURF) [11]. Then, the set of descriptors of the n features found at image I_t is defined as $F_t = \{f_0^t, f_1^t, \ldots, f_{n-1}^t\}$. These descriptors are compared using the Euclidean distance. The reader is referred to Sect. 2.2.1.2 for further information about SIFT or SURF.

5.3 Map Representation

The main goal of FEATMap is to construct a clean visual representation of the robot environment using a monocular camera while localizing the robot within the map. Since in a real scenario storing all images taken by the camera is impossible, we need to reduce the number of images to handle without missing visually distinct locations of the robot environment. The elements of this subset of images are called *keyframes* [12]. FEATMap is also based on the keyframe concept. In our map, each node represents a keyframe, and each keyframe is represented by its corresponding feature descriptors. Formally, given $I = \{I_0, I_1, \ldots, I_t\}$ as the input sequence of images received up to time t, our topological map at time t is defined as:

$$M_t = (\gamma, \omega, \beta), \qquad (5.1)$$

being γ a graph which encodes the relationships between the keyframes, ω the set of selected keyframes up to time t, and β an index of features based on a set of randomized kd-trees that contains the descriptors of the keyframes. Initially, the index is created empty, and it grows as images are received and processed. We empirically found that using 4 trees in parallel is a good compromise between performance and efficiency. More precisely, ω is defined as:

$$\omega = \{\kappa_0, \kappa_1, \ldots, \kappa_{c-1}\}, \qquad (5.2)$$

where κ_i is the keyframe i from a total of c selected keyframes. Each keyframe κ_i is represented by the set of descriptors F_j found in the corresponding image I_j.

The index β is a key component used during the loop closure detection step. As we will see shortly, it is needed to match efficiently the features of the current image with features of all previously considered keyframes, in order to determine whether it is a revisited place. Therefore, a method for an efficient nearest neighbour search is required in order to match these high-dimensional descriptors. Tree structures have been widely used to this end, since they reduce the search complexity from linear to logarithmic. To the same purpose, we maintain a set of randomized kd-trees containing all the descriptors of the detected keyframes. An inverted index structure, which maps each feature to the keyframe where it was found, is also employed. Given a query descriptor, these structures allow us to obtain, traversing the tree just once, the top K nearest keypoints among all keyframes in an efficient way.

5.4 Topological Mapping Framework

5.4.1 Algorithm Overview

FEATMap is outlined in Fig. 5.1 and Algorithm 5.1. At each time step, the robot is considered to be at a keyframe κ_a of the map. In order to select the keyframes, we discard: (a) images similar to the current keyframe of the robot, since they do not provide distinct visual information about the environment and therefore are redundant; and (b) robot camera turns, because they are noisy and can introduce errors in the mapping and localization processes. For the first case, feature descriptors of the current image are matched applying the ratio test [10] to the features of the current keyframe κ_a:

$$\frac{d_f(f_i^t, f_m^a)}{d_f(f_i^t, f_n^a)} < \rho, \qquad (5.3)$$

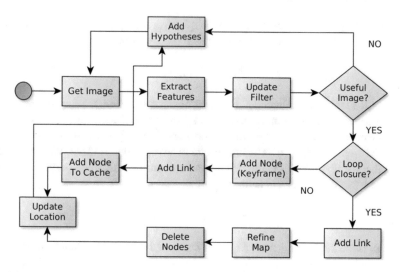

Fig. 5.1 Overview of FEATMap

Algorithm 5.1 FEATMap: Topological Mapping Framework

1: **procedure** TOPOLOGICAL_MAPPING
2:    **while** there are images **do**
3:        I_t = get_image()
4:        F_t = local_description(I_t)   ▷ Feature detection/description
5:        bayes_filter_predict()
6:        likelihood = compute_likelihood(F_t, M_t)
7:        bayes_filter_update(likelihood)
8:        **if** useful_image(F_t, F_{t-1}, F_{k_a}) **then**
9:            **if** loop_closure(F_t, M_t) **then**
10:               κ_c = get_loop_closure_location()
11:               link(κ_a, κ_c, M_t)
12:               refine_map(M_t, κ_c)   ▷ Starts the map refinement step
13:               $\kappa_a = \kappa_c$ ▷ Updates the active location
14:           **else**
15:               κ_n = create_new_location(M_t)
16:               link(κ_a, κ_n, M_t)
17:               $\kappa_a = \kappa_n$ ▷ Updates the active location
18:           **end if**
19:       **end if**
20:       add_hypotheses(t)   ▷ Add valid hypotheses at time t
21:   **end while**
22: **end procedure**

---

being f_i^t a descriptor of the current image I_t, f_m^a and f_n^a, respectively, the nearest and the second-nearest neighbours of the descriptor f_i^t in the keyframe κ_a, d_f the distance between two descriptors and ρ the desired ratio. If the number of matched features is high enough, the image is considered similar to the current location. The same matching step is applied between the current image I_t and the last received

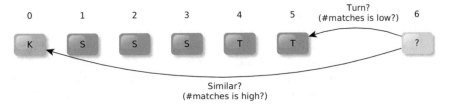

Fig. 5.2 Image selection policy. The current image taken by the camera (6) is matched with the image that represents the current keyframe (0) and the last received image in the sequence (5) in order to determine if it is a useful frame. K represents the current location (keyframe), S and T represent images discarded because they are, respectively, similar enough to the current location or they correspond to camera turns

image in the sequence I_{t-1}: if it is not possible to match a certain number of features, the image is classified as a turn. In these two cases, the image is discarded. Otherwise, it is considered useful and is processed in order to determine whether it is a loop closure or a new keyframe to be added to the map. This keyframe selection policy is shown graphically in Fig. 5.2.

Our loop closure approach makes use of a discrete Bayes filter. This filter is updated with every image irrespective of whether the image has been discarded or not. When an image is considered as useful, FEATMap validates whether it represents a loop closure. If this is not the case, the current image is considered as a new keyframe and is added to the map as a new node. Otherwise, a link is created between the current keyframe and the loop closure candidate and, then, a map refinement process runs, in order to determine if redundant paths have been created. As a consequence, a set of superfluous nodes may be detected. If this is the case, they are removed from the map, and the robot position within the map is updated accordingly.

In order to avoid false loop closure detections between the current image and its neighbours in the sequence, new keyframes are not inserted directly as loop closure hypotheses in the filter. They are instead stored in a temporary cache list and pushed into the filter once a certain number of images have been considered.

The loop closure detection algorithm and the map refinement strategy are detailed in the following sections.

5.4.2 Probabilistic Loop Closure Detection

A discrete Bayes filter is used to detect loop closure candidates. This filter estimates the probability that the current image closes a loop with previously seen locations, allowing us to deal with noisy measurements and uncertainty in the robot location, and helping us to discard false recognitions. The Bayesian framework is also used for ensuring temporal coherency between consecutive predictions, integrating past estimations over time. It could also be used for fusing sensory information from different sources, such as cameras, lasers or IMUs, providing an observation model for each one. In our approach, we only use images as input.

Given the current image I_t at time t, we denote z_t as the observation in our filter, which in this case corresponds to F_t, the set of descriptors extracted from I_t. We also denote L_i^t as the event that image I_t closes a loop with image I_i, where $i < t$. Using these definitions, we want to detect the image of the map I_c whose index satisfies:

$$c = \arg\max_{i=0,\ldots,t-p} \{P\left(L_i^t | z_{0:t}\right)\}, \tag{5.4}$$

where $P\left(L_i^t | z_{0:t}\right)$ is the full posterior probability at time t given all previous observations up to time t. As in [13], the most recent p images are not included as hypotheses in the computation of the posterior since I_t is expected to be very similar to its neighbours and then false loop closure detections will be found. This parameter p delays the publication of hypotheses and needs to be set according to the frame rate or the velocity of the camera.

Separating the current observation from the previous ones, the posterior can be rewritten as:

$$P\left(L_i^t | z_{0:t}\right) = P\left(L_i^t | z_t, z_{0:t-1}\right), \tag{5.5}$$

and then, using conditional probability properties,[1] the next equality holds:

$$P\left(L_i^t | z_t, z_{0:t-1}\right) P\left(z_t | z_{0:t-1}\right) = P\left(z_t | L_i^t, z_{0:t-1}\right) P\left(L_i^t | z_{0:t-1}\right), \tag{5.6}$$

from where we can isolate our final goal to obtain:

$$P\left(L_i^t | z_t, z_{0:t-1}\right) = \frac{P\left(z_t | L_i^t, z_{0:t-1}\right) P\left(L_i^t | z_{0:t-1}\right)}{P\left(z_t | z_{0:t-1}\right)}. \tag{5.7}$$

where $P\left(z_t | z_{0:t-1}\right)$ can be seen as a normalizing factor since its computation does not depend on L_i^t. Under this premise and the Markov assumption, the posterior is defined as:

$$P\left(L_i^t | z_{0:t}\right) = \eta \, P\left(z_t | L_i^t\right) P\left(L_i^t | z_{0:t-1}\right), \tag{5.8}$$

where η represents the normalizing factor, $P\left(z_t | L_i^t\right)$ is the observation likelihood and $P\left(L_i^t | z_{0:t-1}\right)$ is the probability distribution after a prediction step. Decomposing the right side of Eq. 5.8 using the Law of Total Probability, the full posterior can be written as:

$$P\left(L_i^t | z_{0:t}\right) = \eta \, P\left(z_t | L_i^t\right) \sum_{j=0}^{t-p} P\left(L_i^t | L_j^{t-1}\right) P\left(L_j^{t-1} | z_{0:t-1}\right), \tag{5.9}$$

---

[1] $P(A \mid B, C) P(B \mid C) = \frac{P(A \cap B \cap C)}{P(B \cap C)} \frac{P(B \cap C)}{P(C)} = \frac{P(A \cap B \cap C)}{P(A \cap C)} \frac{P(A \cap C)}{P(C)} = P(B \mid A, C) P(A \mid C).$

where $P\left(L_j^{t-1}|z_{0:t-1}\right)$ is the posterior distribution computed at the previous time instant and $P\left(L_i^t|L_j^{t-1}\right)$ is the transition model.

Unlike Angeli et al. [13] and Cummins and Newman [14], we do not model explicitly the probability of no loop closure in the posterior. If the loop closure probability of I_t with I_c ($P\left(L_c^t|z_{0:t}\right)$) is not high enough, we discard L_c^t as a possible loop candidate.

5.4.2.1 Transition Model

Before updating the filter using the current observation, the loop closure probability at time t is predicted from $P\left(L_j^{t-1}|z_{0:t-1}\right)$ according to an evolution model. The probability of loop closure with an image I_j at time $t-1$ is diffused over its neighbours following a discretized Gaussian-like function centred at j. In more detail, 90% of the total probability is distributed among j and exactly four of its neighbours ($j-2$, $j-1$, j, $j+1$, $j+2$) using coefficients $(0.1, 0.2, 0.4, 0.2, 0.1)$, i.e. $0.9 \times (0.1, 0.2, 0.4, 0.2, 0.1)$. The remaining 10% is shared uniformly across the rest of loop closure hypotheses according to:

$$\frac{0.1}{\max\{0, t - p - 5\} + 1}. \tag{5.10}$$

This implies that there is always a small probability of jumping between hypotheses far away in time, improving the sensitivity of the filter when the robot revisits old places.

Our model is similar to the one presented by Angeli [13] but using different coefficients in order to give more importance to the central image of the Gaussian. FAB-MAP employs a Gaussian-like function using only two neighbours, reducing the speed transition of the filter.

5.4.2.2 Observation Model

Once the prediction step is performed, the current observation needs to be included in the filter. We have to compute the most likely locations given the current image I_t and its feature descriptors F_t, but we want to avoid comparing I_t with each previous keyframe, since this can be unfeasible for a high number of images. To this end, we use the index β described in Sect. 5.3. Note that if the robot revisits the same place several times and the current image I_t closes this loop again, each descriptor in F_t can be close to descriptors from different previous images. This fact is taken into account in the computation of our likelihood.

Since we do not use a BoW model, we can not rely on solutions created for these representations like the TF-IDF score [15] used by Angeli [13], or on an observation likelihood based on a precomputed Chow–Liu tree like Cummins [14]. Instead, our observation model provides an efficient way of obtaining loop closure candidates using local scale invariant features and indexing structures such as kd-trees. For each hypothesis i in the filter, a score $s(F_t, F_i)$ is computed. This score represents the likelihood that the current image I_t closes the loop with image I_i given their descriptors, F_t and F_i, respectively. Initially, these scores are set to 0 for all frames from 0 to $t - p$. For each descriptor in F_t, the K closest descriptors among the previous keyframe images are retrieved without taking into account the p previous frames; next, each of them, denoted by n, adds a weight w_n to the score of the image where it appears. This value is normalized using the total distance of the K candidates retrieved:

$$w_n = 1 - \frac{d_n}{\sum_{k=1}^{K} d_k} , \forall n = 1, \ldots, K , \tag{5.11}$$

where d is the Euclidean distance between the considered query descriptor in F_t and the nearest neighbour descriptor found in the tree structure. This value is accumulated onto a score according to:

$$s\left(F_t, F_{j(n)}\right) = s\left(F_t, F_{j(n)}\right) + w_n , \forall n = 1, \ldots, K , \tag{5.12}$$

being $j(n)$ the index of the image where the candidate descriptor n was extracted. The computation of the scores finishes when all descriptors in F_t have been processed. Then, the likelihood function is calculated according to the following rule (similarly to [13]):

$$P\left(z_t | L_i^t\right) = \begin{cases} \dfrac{s(F_t, F_i) - s_\sigma}{s_\mu} & \text{if } s(F_t, F_i) \geq s_\mu + s_\sigma \\ 1 & \text{otherwise} \end{cases} , \tag{5.13}$$

being respectively s_μ and s_σ the mean and the standard deviation of the set of scores. Notice that by means of Eq. 5.13, given the current observation z_t, only the most likely locations update their posterior. After incorporating the observation into our filter, the full posterior is normalized in order to obtain a probability distribution.

Our observation model enables us to detect similar past scenes in challenging situations such as illumination changes, appearance modifications, camera rotations and scene occlusions. This will be shown empirically in the experimental results section, where the observation likelihood for this kind of loop closure situations presents clear peaks despite their complexity. Note that despite the loop closure algorithm is defined for all previous images from 0 to $t - p$, only the likelihood of the images selected as keyframes is computed, since the index β includes only the features of these images. Then, only keyframe images will be returned by the Bayes filter as loop closure candidates.

Algorithm 5.2 FEATMap: Loop Closure Detection

1: **procedure** LOOP_CLOSURE(F_t, M_t)
2:     $[c, P_c]$ = get_best_candidate()   ▷ P_c: Sum of probabilities for candidate c
3:     **if** $P_c > \tau_{loop}$ and number_of_hyp $> \tau_{hyp}$ **then**
4:        ninliers = epipolar_geometry(F_t, F_c)
5:        **if** ninliers $> \tau_{ep}$ **then**
6:           **return** true   ▷ Loop closure found
7:        **else**
8:           **return** false   ▷ Loop closure rejected
9:        **end if**
10:   **else**
11:      **return** false   ▷ No loop closure found
12:   **end if**
13: **end procedure**

5.4.2.3 Selection of a Loop Closure Candidate

In order to select a final candidate, we do not search for high peaks in the posterior distribution, because loop closure probabilities are usually diffused between neighbouring images. This is due to visual similarities between consecutive keyframes in the sequence. Instead, for each location in the filter, we sum the probabilities along a predefined neighbourhood. This neighbourhood is the same as defined in Sect. 5.4.2.1, i.e. frames $(j - 2, j - 1, j, j + 1, j + 2)$ for image j.

The image I_j with the highest sum of probabilities in its neighbourhood is selected as a loop closure candidate. If this probability is below a threshold τ_{loop}, the loop closure hypothesis is not accepted. Otherwise, an epipolarity analysis between I_t and I_j is performed in order to validate if they can come from the same scene after a camera rotation and/or translation. Matchings that do not fulfil the epipolar constraint are discarded by means of RANSAC. If the number of surviving matchings is above a threshold τ_{ep}, the loop closure hypothesis is accepted; otherwise, it is definitely rejected.

Finally, we define another threshold τ_{hyp} to ensure a minimum number of hypotheses in the filter, so that loop closure candidates are meaningful. This step counteracts the fact that first images inserted in the filter tend to attain a high probability of loop closure after the normalization step, what leads to incorrect detections.

The full loop closure detection approach is outlined in Algorithm 5.2.

5.4.3 Map Refinement

Visual topological maps tend to contain redundant nodes and paths due to several reasons. On the one hand, sometimes the current image acquired by the robot is blurred, what makes difficult to identify loop closures at the right time and therefore new nodes are added to the map. The loop closure is identified once the image stream becomes stable again. The net result is that a redundant path is generated because of the noisy images. On the other hand, the Bayes filter does not detect a revisited place instantaneously, but needs some frames to become aware of the loop closure: along these frames, the posterior moves from one keyframe (hypothesis) to another, while a new path containing the unmatched frames is created. These problems are common of many vision-based topological mapping solutions. In this section, we present a map refinement framework based on the visual information obtained from each node of the environment in order to maintain the map structure as simple as possible in storage and computational terms.

Our method is executed each time a loop closure is detected. The idea is to refine the local area of the map around the loop closing node, since the redundant paths are generated within this zone. To this end, its k-neighbourhood is obtained. This is the set of nodes from which we can reach the loop closing node in k steps or less, where k was set empirically to 10. For each element in this set, all paths to the loop closing node are obtained using an adjacency list. If there is only one path between the nodes, it is concluded that there are no redundant paths and this route is left unaltered. Otherwise, a further analysis of the different paths is performed. To this end, a path P of length l between nodes i and j is defined as:

$$P_j^i = \{N_0, N_1, \ldots, N_l\}, 0 \le l \le k+1, \tag{5.14}$$

being N_0 the starting node of P_j^i and N_l the loop closing node. We define the *erasability* of a path as:

$$E(P_j^i) = (deg^-(N_i) = 1) \wedge (deg^+(N_i) = 1), \forall i = 1, \ldots, l-1, \tag{5.15}$$

where deg^- and deg^+ are, respectively, the input degree and the output degree of a vertex. Therefore, a path is classified as *erasable* if Eq. 5.15 holds for each inner node of the path. Otherwise, the path is classified as *non-erasable*. The meaning of an erasable path in our context is that the route can be deleted without breaking the topology of the environment. Examples of erasable and non-erasable paths are shown in Fig. 5.3.

Once all paths have been classified according to their erasability, a decision about which ones can be deleted is made, taking into account that real alternative paths have to be preserved to maintain the topology of the environment. To this end, we propose to generate a reference path using the feature descriptors of all paths, in order to summarize the visual appearance of the route. Next, each erasable path is contrasted against this reference to validate if it is a redundant or a required path. In order to create the reference path, a k-means clustering process with 100 centroids is performed using the descriptors of the keyframes that belong to all paths in consideration, resulting into a set of reference virtual descriptors. Next, for each erasable path, the same clustering process is applied using the descriptors of the corresponding nodes of the map. The virtual descriptors of each path are matched against the reference path using a brute-force approach. Then, the distance between the paths is defined as the average distance between the matched descriptors.

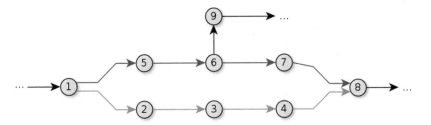

Fig. 5.3 Example of erasability. Two paths exist between node 1 and node 8. The red path (up) is classified as *non-erasable*, since the inner node 6 does not fulfil Eq. 5.15. This path can not be removed without losing the path starting at node 9. The green path (bottom) is classified as *erasable* and is a candidate to be removed

---

Algorithm 5.3 FEATMap: Map Refinement

1: **procedure** REFINE_MAP(M_t, κ_c)
2:   neighbours = get_neighbours(M_t, κ_c, 10)
3:   **for all** neighbour κ in neighbours **do**
4:     P = get_paths(κ, κ_c)
5:     **if** length(P) > 1 **then**
6:       $E = []$ ▷ Erasability of each path
7:       $D = []$ ▷ Distance of each path to the model
8:       d_r = compute_reference_path(P)
9:       **for each** p in index_of(P) **do**
10:         $E[p]$ = compute_erasability($P[p]$)
11:         d_p = compute_path_descriptor($P[p]$)
12:         $D[p]$ = compute_distance(d_r, d_p)
13:       **end for**
14:       m = max_path(D)   ▷ Gets the index of the farthest path
15:       **for each** p in index_of(P) **do**
16:         **if** $E[p]$ and $D[p] < \tau_{path}$ and (exist_non_erasable(E) or m $\neq$ p) **then**
17:           delete_path($P[p]$, M_t)
18:         **end if**
19:       **end for**
20:     **end if**
21:   **end for**
22: **end procedure**

For each erasable path between the nodes, if the distance is below a threshold τ_{path}, this path is considered similar to the others and thus is regarded as redundant. If there is at least one non-erasable route between the nodes, the inner nodes belonging to the rest of erasable paths are deleted. Otherwise, if all paths are classified as erasable, the most different path (higher distance) is left unaltered, and the remaining ones are removed. The full algorithm for map refinement is outlined in Algorithm 5.3.

Figure 5.4 shows several examples of situations that our map refinement strategy is able to overcome. In (a), (b) and (c), the removed paths were selected because the distances to the model path are lower than the others. In (d), since there exists a non-erasable path between nodes 1 and 3 and the distances of the other paths to the

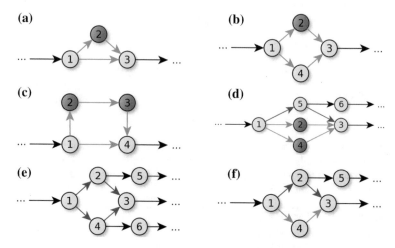

Fig. 5.4 Examples of situations solved by our map refinement strategy. Green and red paths are, respectively, *erasable* and *non-erasable* paths. Red nodes indicate that they will be removed by our approach. When there are several erasable paths, the decision is taken according to the distance of the paths to the reference model path, as explained in the text

model are lower than the threshold τ_{path}, both erasable routes are deleted. In (e), any of the routes can be deleted since they are non-erasable paths. In (f), there exists a non-erasable path and then, the erasable one could be deleted. However, in this case the path can not be removed, since the distance of the path to the model is higher than τ_{path}, indicating that this is a real alternative path.

5.5   Experimental Results

In this section, we will report about the results of several experiments, assessing FEATMap from different points of view. This section is organized as follows: first, we discuss the configuration of the different parameters and their effect in the algorithm; next, the loop closure detection algorithm is evaluated irrespective of the mapping and localization process; then, results for the full mapping and localization approach are shown; finally, experiments for validating our map refinement algorithm are reported.

5.5.1   Parameter Configuration

In all cases, the algorithm was configured using the parameter values indicated in Table 5.1, what achieve the best performance in all sequences used to validate the

Table 5.1 Parameters for
FEATMap execution

Parameter	Value
Keypoints per image (n)	650
Nearest neighbour ratio (ρ)	0.8
Previous images discarded (p)	30
Map refinement neighbourhood (k)	10
Sum of probabilities (τ_{loop})	Varying
Number of inliers for loop closure(τ_{ep})	45
Minimum number of hypotheses (τ_{hyp})	20

approach. In this section, we discuss the election of the most important parameters
and how each one affects the global performance of the algorithm. Particularly:

- The number of keypoints per image (n) has a great impact in the general performance of FEATMap: the higher the number of descriptors to manage by the
  feature index, the higher the time needed to detect a loop closure. However, a low
  number of features per image decreases the ability of the algorithm to find correct
  loop closures, and, as a consequence, increases the number of false positives. We
  found that 650 features per image is enough to avoid false positives in all cases.
- According to Lowe [10], a nearest neighbour ratio (ρ) of 0.8 is enough to eliminate
  90% of the false matches while discarding less than 5% of the correct matches.
- The number of previous discarded images (p) is closely related to the velocity and
  the frame rate of the camera. According to our experiments, a value of 30 ensures
  that no loop closures will be found with the most recent frames, assuming that the
  camera keeps moving.
- The number of inliers after passing the epipolar test to accept a loop closure (τ_{ep})
  affects the sensitivity of the approach: the lower the value, the higher the number
  of loop closures accepted. However, false positives are more likely to appear. Note
  that this parameter is closely related with the number of keypoints per image (n).
  Given that we established n to 650, we found that a value of 45 is enough to avoid
  false positives in all cases.
- The loop acceptance threshold (τ_{loop}) also affects the sensitivity of the algorithm
  and has been used for plotting the precision-recall curves shown in the next section.
  Due to this reason, its value changes from one execution to another of the algorithm.

The remaining parameters were not deemed as critical for the performance of
FEATMap. They were established empirically.

5.5.2 Loop Closure Detection

Several experiments were carried out in order to validate the suitability of FEATMap
for loop closure detection tasks. We processed sequences from indoor and outdoor

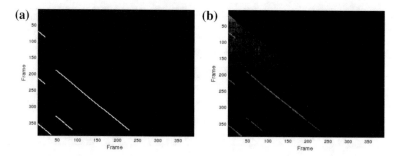

Fig. 5.5 a Ground truth loop closure matrix for the Lip6Indoor sequence. **b** Likelihood matrix computed using FEATMap

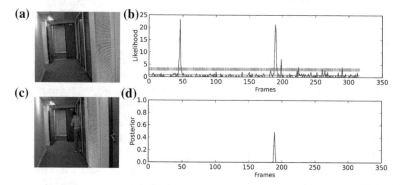

Fig. 5.6 Example of loop closure detection visiting several times the same place and with changes in the environment in the Lip6Indoor sequence. Image 331 (**a**) closes a loop with image 189 (**c**) and image 48 (not shown). As can be seen in **b**, the current likelihood presents two strong peaks despite a person in the current image occludes part of the scene. Peaks correspond to loop candidates. After the prediction, update and normalization steps, the posterior **d** shows a single peak in the last candidate frame. Red and green lines show respectively s_μ and $s_\mu + s_\sigma$ values

environments, providing results under different environmental conditions. More precisely, FEATMap is validated against the Lip6 Indoor, Lip6 Outdoor, UIBSmallLoop, UIBLargeLoop and UIBIndoor sequences. The reader is referred to Sect. 4.2 for further details about them.

Figure 5.5 illustrates the performance of the observation likelihood for detecting loop closures within the Lip6Indoor sequence. The right picture shows the likelihood function values for every pair of frames I_i and I_j while the left picture is the ground truth. As can be seen, our likelihood presents high values for real loop closures, which are shown as diagonals in the images. There is more noise in the likelihood at the beginning of the sequence because there are less images in the trees, which implies that nearest neighbours for each descriptor are shared between a minor number of images. This effect decreases as we move forward along the sequence.

Figure 5.6 shows the performance of the Bayes framework in a loop closure detection situation. In this case, the camera visited twice the same place. When it returns to

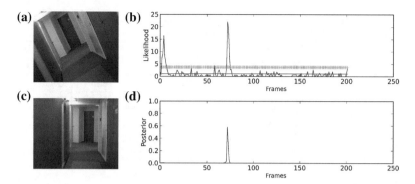

Fig. 5.7 Example of loop closure detection under camera rotations. Despite there is a camera rotation, image 216 (**a**) closes a loop with image 72 (**c**). The likelihood **b** presents two high peaks since it is the third time the camera visits this place. **d** shows the final posterior. Red and green lines show respectively s_μ and $s_\mu + s_\sigma$ values

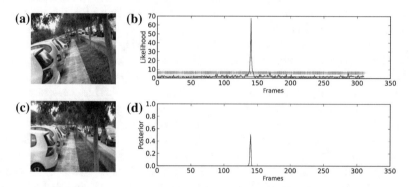

Fig. 5.8 Example of loop closure detection under bad weather conditions and camera rotations for the UIBSmallLoop sequence. Image 330 (**a**) closes a loop with image 139 (**c**). **b** Likelihood given the current image. **d** Full posterior after the prediction, update and normalization steps. Red and green lines show respectively s_μ and $s_\mu + s_\sigma$ values

this place again, two high peaks corresponding to the previous visits can be observed in the likelihood, representing possible loop candidates for the current image. After the prediction, update and normalization steps, the posterior presents only one single peak at the most recent candidate image, corresponding to the closest in time, i.e. the filter ensures temporal coherency between predictions. This figure also shows an example of situation where a loop is detected despite there is a person in the image who was not in the previous visit, what suggests the ability of the filter for detecting loops when the appearance of the environment changes. FEATMap accepts the loop closure since the epipolar constraint between the two images is satisfied. It is also able to detect loop closures under camera rotations, as can be seen in Fig. 5.7, and under bad weather conditions, as shown in Fig. 5.8.

Table 5.2 Results for the five sequences using FEATMap and FAB-MAP 2.0. Precision (Pr) and Recall (Re) columns are expressed as percentages

Sequence	FEATMap						FABMAPv2	
	TP	TN	FP	FN	Pr	Re	Pr	Re
Lip6Indoor	191	151	0	31	**100**	**86.04**	33.12	66.04
Lip6Outdoor	551	435	0	52	**100**	**91.38**	100	13.26
UIBSmallLoop	194	172	0	2	**100**	**99.97**	100	28.84
UIBLargeLoop	439	491	0	47	**100**	**90.32**	100	19.14
UIBIndoor	157	177	0	30	**100**	**83.95**	88.34	8.05

If e.g. an overexposed image or with not enough features was considered by the filter, the full posterior might not present high peaks and a false negative could be generated. However, as soon as the image stream became stable, the algorithm would react and start detecting loop closures again. This shows that FEATMap is able to manage challenging situations.

For obtaining global performance measures, FEATMap was assessed in terms of precision-recall, as explained in Chap. 4. The results for each sequence are shown in Fig. 5.9 and Table 5.2. The best recall rates for 100% precision are shown in the table. As can be seen, no false positives resulted in any case. This is essential, since false positives can induce errors in mapping and localization tasks.

In order to validate the reliability of our loop closure algorithm against other existing solutions, we performed a comparison with the state-of-the-art FAB-MAP 2.0 algorithm [16], whose binaries and visual vocabularies for indoors and outdoors are available online. The output of the algorithm is processed as explained in Sect. 4.3.1. The curves for FEATMap result from modifying the threshold for loop acceptance (τ_{loop}). Clearly, our approach outperforms FAB-MAP in all sequences, obtaining a higher recall for 100% precision. As can be seen, our solution is also more stable, specially in indoor environments, where the performance of FAB-MAP decreases dramatically. We think this is due to the use of the indoors vocabulary and the complexity of finding features in this kind of environment. As a further benefit, our approach can deal better with sensor noise. Notice that a precision below 100% implies the presence of false positives, what will have an influence over the generated map. Our algorithm allows us to obtain a higher recall than FAB-MAP maintaining the maximum precision possible. The maximum precision of FAB-MAP together with its recall for each sequence is also shown in Table 5.2.

As can be seen, a high rate of correct detections were obtained for all experiments. False negatives are due to, on the one hand, the sensitivity of the filter. In effect, when an old place is revisited, the likelihood associated to that hypothesis needs to be higher than the other likelihood values during several consecutive images in order to increase the posterior for this hypothesis. This introduces a delay in the loop closure detection, which derives in false negatives. This sensitivity can be tuned by modifying the transition model of the filter, although a higher sensitivity can

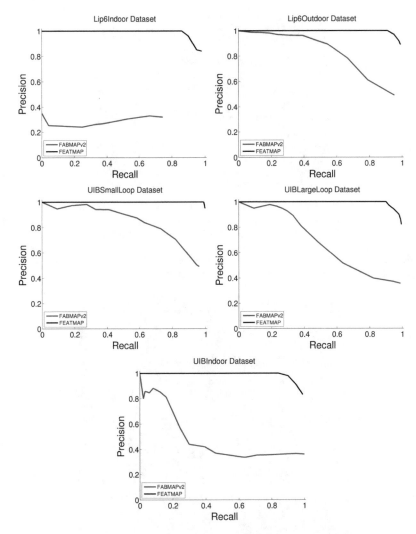

Fig. 5.9 Precision-recall curves for each sequence using FEATMap and FAB-MAP 2.0

introduce loop detection errors, i.e. false positives. On the other hand, false negatives can also be due to camera rotations. When the camera is turning around a corner, it is difficult to find and match features in the images, which prevents the hypothesis from satisfying the epipolar constraint and leads to the loop closure hypothesis to be rejected, despite the posterior for this image is higher than τ_{loop}. However, in spite of the difficulties of the UIBIndoor sequence, our approach is able to succeed, as can be seen in Table 5.2.

The paths followed by the camera in the UIB sequences are shown in Figs. 5.10 and 5.11. These sequences do not include a file with image poses and, then, the

Fig. 5.10 Path followed by the camera during the UIBSmallLoop experiment. Green and blue points indicate respectively the beginning and the end of the sequence; the black lines show no loop closure detections (highest posterior probability is under T_{loop}) and the yellow lines represent loop closure detections (highest probability is above T_{loop} and the epipolar constraint is satisfied). Notice that the camera passes through the same place in successive loops, but the lines are drawn in parallel for visualization purposes. Original image taken from Google Maps

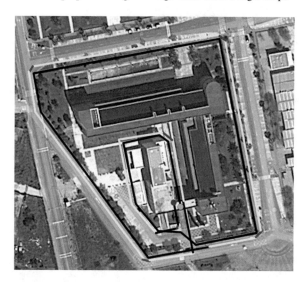

Fig. 5.11 Path followed by the camera during the UIBLargeLoop experiment. Green and blue points indicate respectively the beginning and the end of the sequence; the black lines show no loop closure detections (highest posterior probability is under T_{loop}), the red lines show rejected hypotheses (no epipolar geometry is satisfied) and the yellow lines represent loop closure detections (highest probability is above T_{loop} and the epipolar constraint is satisfied). Notice that the camera passes through the same place in successive loops, but the lines are drawn in parallel for visualization purposes. Original image taken from Google Maps

positions are approximately plotted to show the loop closures detected. Whenever the camera explores new places, no loop closures are found. When a place is revisited, the algorithm starts to find loop closures. Several images are usually needed until closing the loop, due to the filter inertia. These images correspond to the false negatives found.

5.5.3 Topological Mapping and Localization

The same sequences used in the previous experiments were also employed to validate our framework regarding mapping and localization. To this end, the loop closure detection algorithm was adapted to be used with the detected keyframes. A real map of the environment and the topological map generated by our approach are shown for each sequence. The main zones of these maps were labelled with letters to simplify the identification of each part in the topological structure, since these maps do not preserve the shape. The results are shown from Figs. 5.12, 5.13, 5.14, 5.15, 5.16 and 5.17.

As can be seen, the maps generated by FEATMap represent topologically the real scenario. Connections between each part of the topological map are the same of the real environment, and the maps do not contain redundant paths or spurious nodes between locations, saving storage space and improving the computational efficiency of the localization process. Therefore, we can conclude that our map refinement strategy helps us to clean the final structure, correcting the problems generated by blurred images and the delays inherent to the loop closure detection process.

Maps are mainly created during the first exploration of the environment, so that revisiting a place normally turns into reassigning the current location of the robot to an existing node of the map. However, sometimes maps result enlarged with new nodes corresponding to images which are visually in-between two nodes. Generally, they provide unregistered information about the robot scenario, as can be seen for example in Fig. 5.15.

To finish, it must be noted that it is typical that a few nodes at the beginning of the sequence do not close any loop, generating a short tail in the map. This is due to the prediction of the Bayes filter, which tends to move the probability away from the beginning of the sequence, producing that the first loop is closed with the subsequent frames. Notice that this, however, does not affect the final result of the localization process.

5.5.4 Map Refinement

The main goal of this last section is to verify the quality of the refined maps, in terms of storage space, computational times and usefulness/efficiency. We want to assess that the generated maps are representative of the environment and can be used for localization without compromising the original performance. To this end,

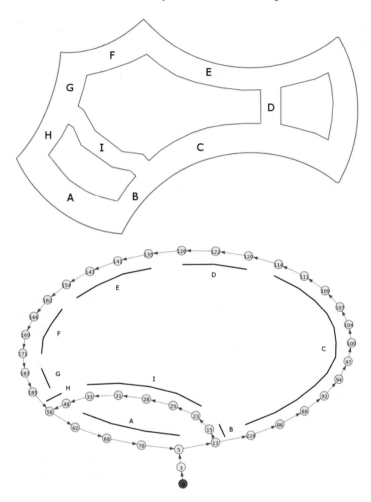

Fig. 5.12 (top) Reference map for the Lip6Indoor sequence. (bottom) Topological map generated using FEATMap. Each part of the map is identified with a letter in both maps. The red node identifies the beginning of the sequence. The original reference map comes from http://cogrob.ensta-paristech. fr/loopclosure.html. Map locations are visited in the following order: A-B-I-H-A-B-C-D-E-F-G-H-A-B-C-D-E-F-G-H-A-B-I

we compare the maps of the five sequences used in this work with and without refinement. The former appear from Figs. 5.12, 5.13, 5.14, 5.15, 5.16 and 5.17, while the latter are shown from Figs. 5.18, 5.19, 5.20, 5.21 and 5.22. As can be seen, the original maps without the refinement contain spurious nodes and alternative redundant paths between nodes, incrementing the execution time of mapping and localization processes, since more nodes need to be considered at each step. The refined maps shown in Sect. 5.5.3 represent better the environment.

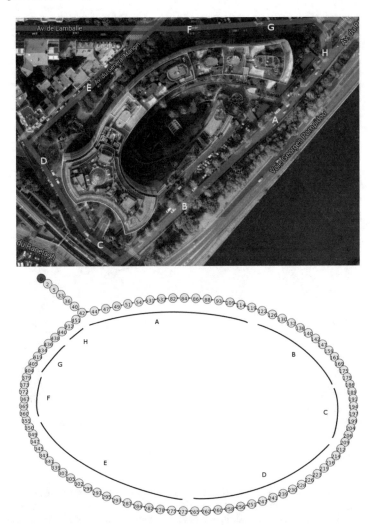

Fig. 5.13 (top) Reference map for the Lip6Outdoor sequence. Original image taken from Google Maps. (bottom) Topological map generated using our approach. Each part of the map is identified with a letter in both maps. The red node identifies the beginning of the sequence. Map locations are visited in the following order: A-B-C-D-E-F-G-H-A-B-C-D-E-F-G-H-A-B

An additional experiment was performed in order to verify whether refined maps could be employed for localization with a similar performance to the original ones. To this end, we first generated the map of the environment and, for each image, the assigned keyframe was stored. After that, the sequence was processed again using the localization filter to determine, for each image, the closest location in the map. If that location was the same as the one stored during the mapping process, the image was considered as a correct localization (CL). For each sequence, we also obtained

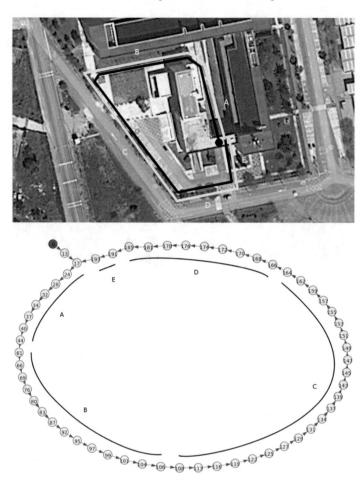

Fig. 5.14 (top) Reference map for the UIBSmallLoop sequence. Original image taken from Google Maps. (bottom) Topological map generated using our approach. Each part of the map is identified with a letter in both maps. The red node identifies the beginning of the sequence. Map locations are visited in the following order: A-B-C-D-E-A-B-C-D-E

Fig. 5.15 Example of adding intermediate nodes in the Lip6Indoor sequence. Images 13 (**a**) and 86 (**c**) were added to the map at the first loop. Image 228 (**b**) was added to the map the next time the camera visited the same place. As can be seen, image 228 is visually in-between the left and the right images

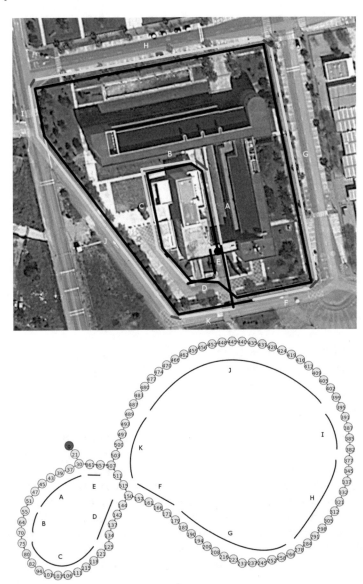

Fig. 5.16 (top) Reference map for the UIBLargeLoop sequence. Original image taken from Google Maps. (bottom) Topological map generated using our approach. Each part of the map is identified with a letter in both maps. The red node identifies the beginning of the sequence. Map locations are visited in the following order: A-B-C-D-F-G-H-I-J-K-F-G-H-I-J-K-E-A-B-C-D-E

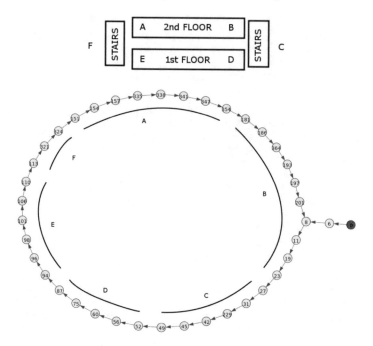

Fig. 5.17 (top) Reference map for the UIBIndoor sequence. (bottom) Topological map generated using our approach. Each part of the map is identified with a letter in both maps. The red node identifies the beginning of the sequence. Map locations are visited in the following order: B-C-D-E-F-A-B-C-D-E-F-A-B

the total mapping and localization times, as well as the number of nodes generated in the graph. These values were measured for each sequence with and without the refinement step. The results can be found in Table 5.3. As it is shown, map refinement leads to less nodes than without it. Despite the correct localization rate is slightly lower for some environments, refining the map improves the computational times of the mapping and localizations processes. This effect increases with the length of the sequence, as is the case of the Lip6Outdoor and the UIBLargeLoop sequences. The UIBSmallLoop sequence presents small differences between the two versions of the map. This is because the resulting structures in the maps are practically the same, resulting into similar processing times. From the table we can also observe that, in general, the outdoor environments are more affected by the refinement step, since the correct localization rates are lower for these cases. In general terms, we can argue that the map refinement strategy proposed in this work can be used for saving space in memory and for improving the speed of the mapping and localization tasks without compromising the performance of FEATMap.

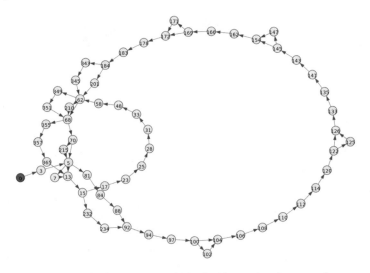

Fig. 5.18 Map of the Lip6Indoor sequence obtained without using the map refinement strategy. The red node identifies the beginning of the sequence

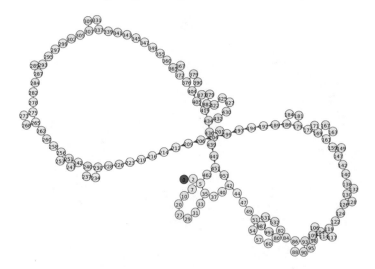

Fig. 5.19 Map of the Lip6Outdoor sequence obtained without using the map refinement strategy. The red node identifies the beginning of the sequence

5.5.5 *Computational Times*

In this section, we evaluate the performance of FEATMap in terms of computational time. To this end, we execute the algorithm over the KITTI 05 sequence. For comparing FEATMap against the solutions that will be presented in next chapters, we assume the worst possible case during the execution of FEATMap, which in this

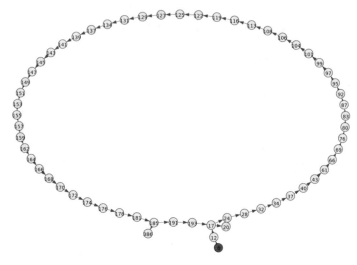

Fig. 5.20 Map of the UIBSmallLoop sequence obtained without using the map refinement strategy. The red node identifies the beginning of the sequence

Fig. 5.21 Map of the UIBLargeLoop sequence obtained without using the map refinement strategy. The red node identifies the beginning of the sequence

case corresponds to insert each image as a new node into the map. The times needed to execute the different parts of the algorithm have been measured, and the results are shown in Fig. 5.23 and summarized in Table 5.4, where *SIFT* and *SURF* are the times needed to describe 1000 features each, *Likelihood Computation* is the time needed to compute the likelihood for the corresponding image, *Bayes Predict* is the time needed to make a prediction in the filter and *Bayes Update* is the time needed

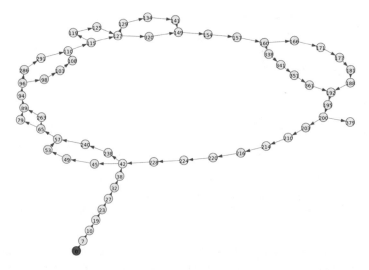

Fig. 5.22 Map of the UIBIndoor sequence obtained without using the map refinement strategy. The red node identifies the beginning of the sequence

Table 5.3 Results for the map refinement experiment. N: number of nodes; M: mapping time in seconds; L: localization time in seconds; $\%CL$: ratio between correct localizations and total number of elements

Sequence	With map refinement				Without map refinement			
	N	M	L	%CL	N	M	L	%CL
Lip6Indoor	40	137.37	6.27	64	62	155.36	9.08	63
Lip6Outdoor	103	1005.52	29.05	63	141	1150.87	42.95	61
UIBSmallLoop	59	152.2	8.25	75	61	155.3	8.57	77
UIBLargeLoop	100	728.94	44.29	74	111	798.53	60.98	78
UIBIndoor	40	118.4	16.2	76	59	190.39	24.64	73

to update the filter using the computed likelihood. The total values of the last row are calculated only summing the cost of computing SURF features, since it is considered as the fastest description technique. The loop closure times correspond to using SIFT descriptors. Given that the length of the SIFT and SURF descriptors is the same, these loop closure times are very similar for both descriptors.

FEATMap can process one image in 1759.41 ms on average according to our results. As can be seen, most part of the time is invested in the image description step, specially when using SIFT, and during the likelihood computation. These are the main bottlenecks of FEATMap. The Bayes filter steps can be considered fast in comparison with these two steps. As shown, the time needed to compute the likelihood increases as more keyframes are inserted in the map, which can be a problem in large environments.

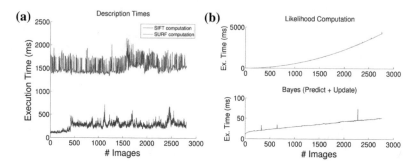

Fig. 5.23 Computational times of FEATMap executed over the KITTI 05 sequence. **a** Computational times for describing an image. **b** Computational times for loop closure detection. Likelihood and Bayes filter computational times have been separated in two different plots for improving their visualization

Table 5.4 Computational times for FEATMap executed over the KITTI 05 sequence. All times are expressed in milliseconds. The totals are computed taking into account the fastest description method, which in this case corresponds to SURF. The % column indicates the percentage of total time that approximately represents each step

		Mean	%	Std	Max	Min
Image description	SIFT	1543.30	–	137.30	2156.60	1324.60
	SURF	283.60	16.12	80.10	827.30	84.40
Loop closure	Likelihood	1439.02	81.79	1284.40	4511.40	0.10
	Bayes predict	34.22	1.94	10.28	73.41	10.44
	Bayes update	2.57	0.15	1.08	4.51	0.67
	Total	1759.41		1286.94	5416.62	95.61

5.6 Discussion

In this chapter, we have introduced FEATMap, a complete appearance-based topological mapping and localization framework based on local invariant features. When a new useful image is acquired, a discrete Bayes filter is used to select a loop closure candidate and decide whether this frame closes a loop or a new node to be added to the map. This probabilistic filter presents a novel observation model based on an efficient matching scheme between the current image and the features of the current nodes in the map, using an index based on a set of randomized kd-trees. As a result, a topological map of the environment is obtained, which represents the scenario of the robot as a graph of keyframes.

Using probabilistic filters for mapping and localization tasks usually produces spurious nodes and redundant paths over the graph. This is due to imperfections in the acquired images and the delays introduced by the filter. A key contribution of FEATMap is a map refinement strategy for solving these problems, producing cleaner maps and saving storage space and computation resources for mapping and localization tasks. This technique is executed each time a loop is closed, and a predefined neighbourhood is refined in each step. The final decision of deleting nodes is taken according to the visual features of each path, avoiding the removal of real paths of the environment.

In order to validate our solution, results from an extensive set of experiments, using datasets from different environments, have been reported. These results show that our mapping and localization approach using a map refinement step can be employed for generating topological maps of the environment which, if they are provided with odometry information, can also be used for navigating in the current scenario in an efficient way. FEATMap has been also compared against the state-of-the-art FAB-MAP 2.0 algorithm, obtaining better performance in all the sequences employed during the experiments.

Despite the good results obtained using FEATMap, several drawbacks have been identified. These drawbacks have guided our efforts in the development of the remaining solutions presented in this book. One of the main problems is the description method, since real-valued descriptors like SIFT or SURF can be considered today slow techniques in comparison with the binary description approaches proposed recently. This kind of descriptors can not be indexed using kd-tree structures, as done in FEATMap, since they can not be averaged. We need then efficient structures for indexing binary features.

Another problem is the scalability of the solution: the performance of FEATMap decreases as more keyframes are inserted into the map. The main reason is that all the detected features of the corresponding image are inserted as new features in the index. A BoW scheme, where the features are quantized according to a reference visual dictionary, can help in these cases. However, we are also interested in avoiding the training step, saving processing time and adapting the visual dictionary to the operating conditions. Taking into account these considerations, in the following chapter we introduce an incremental Bag-of-Binary-Words scheme, which is used within a dense topological mapping framework.

References

1. Sivic, J., Zisserman, A.: Video Google: a text retrieval approach to object matching in videos. In: IEEE International Conference on Computer Vision, pp. 1470–1477 (2003)
2. Zhang, H.: BoRF: loop-closure detection with scale invariant visual features. In: IEEE International Conference on Robotics and Automation, pp. 3125–3130 (2011)
3. Angeli, A., Doncieux, S., Meyer, J.A., Filliat, D.: Real-time visual loop-closure detection. In: IEEE International Conference on Robotics and Automation, pp. 1842–1847 (2008)
4. Booij, O., Terwijn, B., Zivkovic, Z., Krose, B.: Navigation using an appearance based topological map. In: IEEE International Conference on Robotics and Automation, pp. 3927–3932 (2007)
5. Calonder, M., Lepetit, V., Strecha, C., Fua, P.: BRIEF: binary robust independent elementary features. In: European Conference on Computer Vision, Lecture Notes in Computer Science, vol. 6314, pp. 778–792 (2010)

6. Leutenegger, S., Chli, M., Siegwart, R.: BRISK: binary robust invariant scalable keypoints. In: IEEE International Conference on Computer Vision, pp. 2548–2555 (2011)
7. Rublee, E., Rabaud, V., Konolige, K., Bradski, G.: ORB: an efficient alternative to SIFT or SURF. IEEE Int. Conf. Comput. Vis. **95**, 2564–2571 (2011)
8. Alahi, A., Ortiz, R., Vandergheynst, P.: FREAK : fast retina keypoint. In: IEEE Conference on Computer Vision and Pattern Recognition, pp. 510–517 (2012)
9. Yang, X., Cheng, K.T.: Local difference binary for ultrafast and distinctive feature description. IEEE Trans. Pattern Anal. Mach. Intell. **36**(1), 188–94 (2014)
10. Lowe, D.G.: Distinctive image features from scale-invariant keypoints. Int. J. Comput. Vision **60**(2), 91–110 (2004)
11. Bay, H., Tuytelaars, T., Van Gool, L.: SURF: speeded up robust features. In: European Conference on Computer Vision, Lecture Notes in Computer Science, vol. 3951, pp. 404–417 (2006)
12. Zhang, H., Li, B., Yang, D.: Keyframe detection for appearance-based visual SLAM. In: IEEE/RSJ International Conference on Intelligent Robots and Systems, pp. 2071–2076 (2010)
13. Angeli, A., Filliat, D., Doncieux, S., Meyer, J.A.: A fast and incremental method for loop-closure detection using bags of visual words. IEEE Trans. Robot. **24**(5), 1027–1037 (2008)
14. Cummins, M., Newman, P.: FAB-MAP: probabilistic localization and mapping in the space of appearance. Int. J. Rob. Res. **27**(6), 647–665 (2008)
15. Sparck Jones, K.: A statistical interpretation of term specificity and its application in retrieval. J. Doc. **28**, 11–21 (1972)
16. Cummins, M., Newman, P.: Appearance-only SLAM at large scale with FAB-MAP 2.0. Int. J. Rob. Res. **30**(9), 1100–1123 (2011)

Chapter 6
Loop Closure Detection Using Incremental Bags of Binary Words

Abstract This chapter introduces a novel method for computing a visual vocabulary online. This binary vocabulary, in combination with an inverted file, conforms an index of images called *OBIndex* (Online Binary Image Index), which can be used to efficiently retrieve previously seen places. This chapter also presents a topological mapping algorithm called *BINMap* (Binary Mapping), which makes use of OBIndex as a key component to obtain loop closure candidates during the likelihood computation.

6.1 Overview

As mentioned in previous chapters, many appearance-based algorithms for loop closure detection developed recently [1–4] are based on the Bag-Of-Words (BoW) approach [5, 6]. Most of them generate the dictionary offline, which implies several drawbacks. An alternative is to build the visual dictionary in an incremental manner, while new images are received.

SIFT [7] and SURF [8] are the most commonly used features in BoW schemes, due to their invariance properties to illumination, scale and rotation changes. However, the detection and description of these features are computationally expensive. Recently there has been a growing interest in the use of binary descriptors, such as BRIEF [9], BRISK [10], ORB [11], FREAK [12] or LDB [13]. These features present advantages over real-valued descriptors since they are faster to compute and require less storage space [4]. Binary features are compared using the Hamming distance, which can be efficiently computed by means of a bitwise XOR operation and bit summation. Modern computers provide hardware support for executing these operations quickly.

A BoW scheme based on binary features can be useful to overcome the drawbacks presented by FEATMap. Then, in this chapter, we introduce a method for computing a visual vocabulary online, avoiding the training phase and making use of binary descriptors. This binary vocabulary, in combination with an inverted file, conforms an index of images that we call *OBIndex* (Online Binary Image Index).[1] Next,

---

[1]http://github.com/emiliofidalgo/obindex.

© Springer International Publishing AG 2018
E. Garcia-Fidalgo and A. Ortiz, *Methods for Appearance-based Loop Closure Detection*, Springer Tracts in Advanced Robotics 122,
https://doi.org/10.1007/978-3-319-75993-7_6

this index is used in a probabilistic topological mapping framework called *BINMap* (Binary Mapping). BINMap, as well as FEATMap, is based on an appearance-based loop closure detection algorithm, where the index of features is a key component for obtaining similar loop closure candidates during the likelihood computation. BIN-Map is validated using several datasets and compared against the FAB-MAP 2.0 algorithm.

The main innovation presented in this chapter has to do with the incremental Bag-of-Binary-Words approach. This approach is, to the best of our knowledge and according to the works reviewed in Chap. 3, the first attempt of using binary features and a BoW scheme incrementally for loop closure detection. Perhaps the work most related to the one explained here is the approach introduced by Galvez-Lopez and Tardos [14], but their visual dictionary is built offline. More recently, Khan and Wollherr introduced IBuILD [15], an incremental Bag-of-Binary-Words approach for loop closure detection. OBIndex [16] solves the scalability problem of IBuILD by means of hierarchical structures.

The chapter is organized as follows: Sect. 6.2 introduces our incremental Bag-of-Binary-Words approach for indexing images, Sect. 6.3 explains the image description techniques used in BINMap, Sect. 6.4 presents the map representation used in BIN-Map, Sect. 6.5 details the topological mapping framework, Sect. 6.6 reports several experiments in order to validate the approach and, finally, Sect. 6.7 concludes the chapter.

6.2   Incremental Bag-of-Binary-Words

As mentioned above, the key component of BINMap is an index of images which is used for detecting loop closures. This approach is also part of the remaining solutions for retrieving similar images presented in this book. It is because of this reason that OBIndex is detailed in this section before introducing BINMap.

6.2.1   Fast Matching of Binary Features

A linear search is not feasible for searching words in a large visual dictionary. This problem is solved using an approximate matching algorithm, which usually employs hierarchical structures, such as kd-trees or hierarchical k-means trees, to speed up the process. Several randomized kd-trees were used in the previous chapter for indexing features, but these structures are not suitable for binary descriptors since they assume that each dimension of the vector can be continuously averaged. Typical matching approaches for binary descriptors include hashing techniques, e.g. Locality Sensitive Hashing (LSH) [17] or Semantic Hashing [18]. Recently, Muja and Lowe [19] have presented an algorithm for matching binary descriptors based on a hierarchical decomposition of the search space which performs better in comparison with hashing approaches. Furthermore, new descriptors can be easily added in this structure, becoming this solution into an interesting option to be used as an incremental Bag-of-Binary-Words scheme.

In order to build a tree, initially all the input descriptors are clustered by means of a k-medoids algorithm using K centres selected randomly, where K is called the *branching factor*. Note that the centres are selected randomly from the input points instead of trying to minimize the squared error between the centres and the elements of the clusters, which results into a simpler and faster method for building the tree. This process is repeated recursively until the number of leaf nodes in each cluster is below a threshold L, which is called the *maximum leaf size*. The authors also proposed to build multiple trees (T) and using them in parallel during the search for improving the performance of the algorithm.

The search is performed starting from the root until reaching a leaf node and, then, the points contained within this leaf are linearly searched in order to find the closest candidates. A priority queue is employed when searching in several trees in parallel. After a single traverse of each of the trees, the search continues using the closest node stored in the priority queue and continuing the process from there. The search finishes when the number of points examined is above a certain threshold. The performance of this index is directly related to the input parameters: the branching factor, the maximum leaf size, the number of search trees and the maximum number of points to examine. According to the results presented in [19], their approach requires less storage space, scales better compared to LSH and results to be very effective for matching binary descriptors.

6.2.2 Online Binary Image Index

In this section, we introduce Online Binary Image Index (OBIndex), our approach to find similar images in a database given a query image. In OBIndex, we use a modified version of the Muja and Lowe's approach as an incremental visual dictionary, where the descriptors stored in the trees represent the words of the visual dictionary. The hierarchical trees are used in combination with an inverted index, which contains, for each word in the dictionary, a list of images where it was found. This allows us to obtain similar image candidates in an efficient way, as explained below. A simple example of these structures is shown in Fig. 6.1.

Since our approach relies on an incremental visual dictionary based on binary features, an updating policy for combining binary descriptors is needed. Averaging each component of the vector is an option for real-valued descriptors, but it can not be considered for the binary case. To solve this issue, we propose to use a bitwise AND operation. Formally, being B a binary descriptor, we use the following update rule:

$$B_{w_i}^t = B_{w_i}^{t-1} \wedge B_q , \qquad (6.1)$$

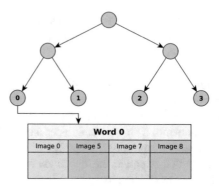

Fig. 6.1 Example of a single hierarchical tree and an inverted file that conform our image index. In this case, the branching factor (K) and the maximum leaf size (L) are set both to 2. The visual vocabulary consists of 4 words $\{w_0, w_1, w_2, w_3\}$. For each of these words, the inverted index stores a list of images in which the word appears

where $B_{w_i}^{t-1}$ is the binary descriptor of the word w_i stored in the dictionary at time instant $t-1$, B_q is the query descriptor and $B_{w_i}^t$ is the merged descriptor stored for the word w_i at time t. This policy is inspired by the observation that each component of a binary descriptor is usually set to 0 or 1 according to the result of a comparison between a pair of image pixel intensities or gradients, e.g. BRIEF, ORB, FREAK and LDB. If the ith bit is the same in both descriptors, it means that the result of this comparison between the pixel intensities was the same in both images. Otherwise, we experimentally prioritize the use of the zero value by means of the AND operation.

6.2.2.1   Adding New Images

The trees are initially built using as visual words the descriptors of the first image to be processed. When an image I_t received at time t needs to be added to the index, its descriptors F_t are searched in the trees. Given a query binary descriptor, we search for the two nearest neighbours traversing the trees from the root to the leafs and selecting at each level the node that minimizes the Hamming distance. Using these two neighbours, we apply the ratio test [7] in order to determine if both descriptors represent the same visual feature. If positive, the query descriptor and the visual word are merged and replaced in the dictionary using Eq. 6.1. Otherwise, the query descriptor is considered a new descriptor and is added to the index as a new visual word, which implies storing the descriptor into the OBIndex structures and adding it to the closest leaf node on each tree. In both situations, the inverted index is updated accordingly, adding a reference to the current image in the list corresponding to the modified or added feature. When the size of the trees is R times bigger than the original one, they are rebuilt to redistribute the descriptors. The updating process of the visual dictionary is summarized in Algorithm 6.1.

Algorithm 6.1 OBIndex: Add New Image

1: **procedure** ADD_NEW_IMAGE(F_t, ρ, β)
2:     F_t: Binary descriptors obtained from image I_t
3:     ρ: Nearest neighbour distance ratio
4:     β: Image index
5:     **for** each d in F_t **do**
6:         $[n1, n2]$ = nearest_neighbours(β, d, 2)   ▷ Two nearest neighbours
7:         **if** distance(d, $n1$) < distance(d, $n2$) * ρ **then**
8:             $B = d \wedge n1$ ▷ Merging descriptors
9:             replace_descriptor(β, $n1$, B)
10:            add_to_inverted_index(β, I_t, $n1$);
11:        **else**
12:            add_new_visual_word(β, d)
13:            add_to_inverted_index(β, I_t, d);
14:        **end if**
15:    **end for**
16: **end procedure**

6.2.2.2 Searching for Images

As mentioned above, OBIndex is a structured database of images that can be used for obtaining, given a query image I_q, an ordered list of similar images. In order to compute this list, OBIndex handles, for each image i stored in the database, a similarity score s_i. Initially, these scores are set to 0 for all images. Given the set of binary descriptors F_q found in the query image I_q, we search each descriptor in the visual dictionary in order to find the closest word. Next, the inverted index allows us to obtain a list of images where this word was found. We then add a statistic about the word to the correspondent score s for each retrieved image. This statistic is inspired in the Term Frequency-Inverse Document Frequency (TF-IDF) weighting factor [20], which reflects how important a word is to the query image I_q with regard to the images received up to time t. It increases the importance of the words seen frequently in a few documents and decreases the importance of the most commonly seen words. Being $I_{0:t}$ the set of the images processed by the index up to time t, the TF-IDF value $\rho^i_{w_j}$ computed given the word w_j and the image I_i is defined as:

$$\rho^i_{w_j} = \mathrm{tf}(w_j, I_i) \times \mathrm{idf}(w_j, I_{0:t}), \tag{6.2}$$

where the term *tf* is the frequency of the word in the image, and the term *idf* is the inverse frequency of the images containing this word. The term *tf* is defined as:

$$\mathrm{tf}(w_j, I_i) = \frac{n^i_{w_j}}{N_i}, \tag{6.3}$$

being $n^i_{w_j}$ the number of occurrences of the word w_j in the image I_i, and N_i the total number of features found in the image I_i. The term *idf* is defined as:

Algorithm 6.2 OBIndex: Search Image

1: **procedure** SEARCH_IMAGE(F_q, β)
2:     F_q: Binary descriptors obtained from query image I_q
3:     β: Image index
4:     $s = []$ ▷ Scores
5:     **for** each image i in β **do**
6:         $s[i] = 0$ ▷ Initializing scores
7:     **end for**
8:     **for** each descriptor d in F_q **do**   ▷ Processing all descriptors
9:         w_j = get_closest_word(β, d)
10:         l = get_images_from_inverted_index(β, w_j)
11:         **for** each image i in l **do**   ▷ Processing each image in the list
12:             $\rho^i_{w_j}$ = compute_tfidf(β, w_j, i);
13:             $s[i] = s[i] + \rho^i_{w_j}$
14:         **end for**
15:     **end for**
16:     sort(s)   ▷ Sorting the candidates in descending order
17:     **return** s
18: **end procedure**

$$\text{idf}(w_j, I_{0:t}) = \log \frac{t-1}{n_{w_j}}, \tag{6.4}$$

where $t-1$ coincides with the cardinal of set $I_{0:t}$, and n_{w_j} is the total number of images in $I_{0:t}$ containing the word w_j. This value is accumulated onto the corresponding score according to:

$$s_i = s_i + \rho^i_{w_j}, \tag{6.5}$$

being i the index of the image extracted from the inverted index. The computation of the scores is finished when all descriptors in F_q have been processed. Then, the list of candidate images are sorted in descending order according to their scores. This search process is outlined in Algorithm 6.2.

6.3 Image Description

Given that OBIndex is descriptor-independent, BINMap can be used with any binary descriptor. This allows us to take advantage of the faster computation and the reduced storage needs of this kind of descriptors, against classic approaches like SIFT [7] or SURF [8]. In BINMap, for each image, we compute a collection of FAST [21] features, and then they are described using the BRIEF [9] algorithm. The detected corners are required to cover the full image in a more or less uniform way. Due to this reason, a 4×4 regular grid is defined over the image and a minimum number of corners in each cell is requested to be found.

Formally, the set of descriptors of the n features found in the image I_t is defined as $F_t = \{f_0^t, f_1^t, \ldots, f_{n-1}^t\}$. The similarity between two binary strings is computed by means of the Hamming distance, which is equal to the number of ones after performing an exclusive OR ($\oplus$) operation between the descriptors. Then, the distance $d_f(f_p^i, f_q^j)$ between two binary descriptors p and q of sets F_i and F_j, respectively, can be defined as:

$$d_f(f_p^i, f_q^j) = \text{bitsum}(f_p^i \oplus f_q^j).   \tag{6.6}$$

6.4 Map Representation

Contrary to FEATMap, BINMap represents the environment as a dense topological map, instead of selecting a set of keyframes from the input images. As was explained previously in Chap. 2, in a dense topological mapping approach, each new image is inserted into the graph as a node. In this case, this map representation is possible thanks to the incremental BoW scheme used in BINMap, which allows us to save computation time and storage space. Unlike FEATMap, where all the features of an image were inserted in the index, the OBIndex quantization process combined with the inverted index results into a more compact representation, improving the scalability of the solution. Then, in BINMap, each node of the graph represents an input image. Formally, given $I = \{I_0, I_1, \ldots, I_t\}$ as the input sequence of images received up to time t, the topological map generated by BINMap is defined as:

$$M_t = (\gamma, \omega, \beta),   \tag{6.7}$$

being γ a graph which encodes the relationships between the images, ω the set of nodes of the graph at time instant t and β an instance of OBIndex, working as explained in Sect. 6.2.2. Given that the number of nodes in BINMap is the same as the number of images received up to time t, in this case w is defined as:

$$\omega = \{n_0, n_1, \ldots, n_t\},   \tag{6.8}$$

where n_i is the node i of the graph, which corresponds to the image I_i. The index β, as in FEATMap, is a key component used during the loop closure detection step. The structures explained in Sect. 6.2.2 are used to find possible loop closure candidates efficiently.

Algorithm 6.3 BINMap: Topological Mapping Framework

1: **procedure** TOPOLOGICAL_MAPPING
2:    **while** there are images **do**
3:       I_t = get_image()
4:       F_t = local_description(I_t)   ▷ FAST and BRIEF computation
5:       n_t = create_new_node(M_t)
6:       link(n_{t-1}, n_t, M_t)   ▷ Linking with the previous node
7:       **if** loop_closure(F_t, M_t) **then**
8:          n_c = get_loop_closure_location()
9:          link(n_t, n_c, M_t)   ▷ Linking with the loop closure node
10:      **end if**
11:   **end while**
12: **end procedure**

6.5 Topological Mapping Framework

6.5.1 Algorithm Overview

The BINMap algorithm is outlined in Fig. 6.2 and Algorithm 6.3. The algorithm is simpler in comparison with FEATMap, given that in this case it is not needed to apply a keyframe selection policy to the input images. For each new image I_t, a set of FAST corners and their correspondent BRIEF descriptors, denoted by F_t, are computed. Next, a new node n_t, corresponding to the current image I_t, is created and linked with the last node inserted in the graph, denoted by n_{t-1}. The BRIEF descriptors are then used in the loop closure detection step in order to determine if this image represents an already visited place. If this is the case, the current node n_t is linked in the graph γ with the loop closure node n_c. Otherwise, no action is performed and the algorithm is ready to process the next image.

The loop closure detection algorithm, which is also based on a probabilistic scheme, is detailed in the following section.

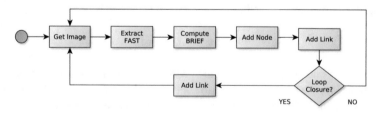

Fig. 6.2 Overview of BINMap

6.5.2 *Probabilistic Loop Closure Detection*

As in FEATMap, a discrete Bayes filter is used to detect loop closure candidates and to integrate measurements over time. Being I_t the current image received at time t, z_t the observation and L_i^t the event that image I_t closes a loop with image I_i, we want to find the image of the map I_c whose index satisfies:

$$c = \arg\max_{i=0,\dots,t-p} \{P\left(L_i^t | z_{0:t}\right)\},$$

where $P\left(L_i^t | z_{0:t}\right)$ is the full posterior probability at time t given all previous observations up to time t. Note that z_t corresponds to the set of binary descriptors extracted from the image I_t, denoted by F_t. Again, we discard the most recent p images as loop closure hypotheses to avoid loop closure detections between neighbouring images. The reader is referred to Sect. 5.4.2 for further details about the derivation of the posterior, since it is the same than the one presented there. Due to this reason, only the final equation is included here for completeness. Consequently, the full posterior can be written as:

$$P\left(L_i^t | z_{0:t}\right) = \eta \, P\left(z_t | L_i^t\right) \sum_{j=0}^{t-p} P\left(L_i^t | L_j^{t-1}\right) P\left(L_j^{t-1} | z_{0:t-1}\right),$$

where η is a normalizing factor, $P\left(z_t | L_i^t\right)$ is the observation likelihood, $P\left(L_i^t | L_j^{t-1}\right)$ is the transition model and $P\left(L_j^{t-1} | z_{0:t-1}\right)$ is the posterior distribution computed at the previous time instant. The observation and transition models, as well as the method for selecting a final loop closure candidate, are explained in the following sections.

6.5.2.1 Transition Model

The loop closure probability at time t is predicted from the previous probability distribution by means of the transition model. In this case, this model is the same as the one used in FEATMap, which was explained in Sect. 5.4.2.1. To summarize, a discretized Gaussian-like function is used to diffuse the 90% of the total probability among four neighbouring images and the image itself. The remaining 10% is shared by the rest of the images by means of the following expression:

$$\frac{0.1}{\max\{0, t - p - 5\} + 1}.$$

6.5.2.2   Observation Model

The observation z_t is incorporated into the filter after the prediction. To this end, the instance of OBIndex, denoted in our map representation by β, is used for an efficient likelihood computation. Note that we delay the publication of hypotheses as loop closure candidates, which implies that the number of images processed by the index β at time t is $t - p$. The current image I_t is queried against the index, resulting into an ordered list of similar images according to their score s_i, as explained in Sect. 6.2.2. The range of these scores depends on the query image and the distribution of the visual words. The resulting list l_t is a sequence of matching candidates:

$$l_t = \{< c_1, s_{c_1} >, < c_2, s_{c_2} >, \ldots, < c_h, s_{c_h} > \mid s_{c_1} > s_{c_2} > \ldots > s_{c_h}\}, \quad (6.9)$$

where c_i is the index of the image candidate and s_{c_i} is its corresponding score. Using this list, the likelihood function is calculated, in a similar way to FEATMap, as:

$$P\left(z_t \mid L_i^t\right) = \begin{cases} \dfrac{s_i - 2s_\sigma}{s_\mu} & \text{if } s_i \geq s_\mu + 2s_\sigma \\ 1 & \text{otherwise} \end{cases}, \quad (6.10)$$

being respectively s_μ and s_σ the mean and the standard deviation of the set of scores. In this case we are more restrictive when selecting the images whose posterior is updated: we require $2s_\sigma$ instead of only s_σ. We have observed that being more restrictive regarding the number of images which are considered for updating their likelihood led to better results. After incorporating the measurement into the filter, the posterior is normalized for obtaining a probability density function.

6.5.2.3   Selection of a Loop Closure Candidate

For selecting a final loop closure candidate, we sum the probabilities along a predefined neighbourhood of each image instead of searching for high peaks, since usually the probabilities are diffused over consecutive images. This is due to the similarities that consecutive images belonging to the same place present. The image I_j with the highest sum of probabilities is analysed in order to determine whether it can close a loop with the current image I_t. If the probability of I_j is below a threshold τ_{loop}, the loop hypothesis is discarded. Otherwise, we check whether the images verify the epipolar geometry constraint by means of a RANSAC-based estimation of the fundamental matrix. If the resulting number of inliers is above a threshold τ_{ep}, the loop closure is accepted; otherwise, it is rejected. As in FEATMap, we also define a threshold τ_{hyp} for avoiding loop closure detections when a small number of images are stored in the filter. The loop closure algorithm is outlined in Algorithm 6.4.

Algorithm 6.4 BINMap: Loop Closure Detection

1: **procedure** LOOP_CLOSURE(F_t, M_t)
2:     ρ: Nearest neighbour distance ratio
3:     enqueue_image(F_t)   ▷ Store the image for inserting it as a future LC candidate
4:     $n = t - p$
5:     **if** $n < 0$ **then**   ▷ Delaying the publication of hypotheses
6:         **return** false
7:     **end if**
8:     add_new_image(F_n, ρ, β_t);   ▷ Add I_n into the index as loop closure candidate
9:     add_hypothesis(n)   ▷ Add I_n as new hypothesis into the Bayes filter
10:     bayes_filter_predict()
11:     likelihood = compute_likelihood(F_t, M_t)
12:     bayes_filter_update(likelihood)
13:     [c, P_c] = get_best_candidate()   ▷ P_c: Sum of probabilities for candidate c
14:     **if** $P_c > \tau_{loop}$ and number_of_hyp $> \tau_{hyp}$ **then**
15:         ninliers = epipolar_geometry(F_t, F_c)
16:         **if** ninliers $> \tau_{ep}$ **then**
17:             **return** true   ▷ Loop closure found
18:         **else**
19:             **return** false   ▷ Loop closure rejected
20:         **end if**
21:     **else**
22:         **return** false   ▷ No loop closure found
23:     **end if**
24: **end procedure**

6.6 Experimental Results

In this section we report several experimental results in order to validate BINMap. This section is organized as follows: first, we discuss the configuration of the different parameters of the algorithm; next, we evaluate the effectiveness of BINMap for detecting loop closures; then, we assess the ability of BINMap for generating topological maps and representing the environment; finally, we discuss computational times.

6.6.1 Parameter Configuration

As in the previous chapter, in this section we discuss how the different parameters affect the performance of BINMap. The parameters were configured using the values indicated in Table 6.1. The parameters K, L, T and R directly affect the performance of OBIndex. Its configuration is highly influenced by the original results obtained by Muja and Lowe [19] and they are detailed next:

- The branching factor (K) has a key impact in the descriptor search performance: the higher the branching factor, the higher the search precision, but the tree build time

Table 6.1 Parameters for BINMap execution

Parameter	Value
Branching factor (K)	16
Maximum leaf size (L)	150
Number of search trees (T)	4
Rebuild threshold (R)	4
Keypoints per image (n)	650
Nearest neighbour ratio (ρ)	0.8
Previous images discarded (p)	30
Sum of probabilities (τ_{loop})	Varying
Number of inliers (τ_{ep})	45
Minimum number of hypotheses (τ_{hyp})	20

is also increased. Muja and Lowe showed that there is a little gain for branching factors above 16 and hence, due to this reason, we set this parameter to 16.

- The effect of the maximum leaf size parameters (L) is evident: the higher this value, the higher the number of descriptors to compare when a leaf node is reached during a search. However, the dispersion of the descriptors stored in the index is reduced, what increases the precision of the search. According to the original results of the authors, a value of 150 is reasonable for this parameter.
- The optimum number of search trees (T) depends on the desired precision. More trees implies more memory and longer tree build times, but also a higher precision. As shown by Muja and Lowe, 4 search trees is a good compromise between precision and speed.
- The rebuild threshold (R) builds a search tree again after a certain number of new descriptors have been inserted. This is useful to reduce the dispersion of the index, but, as more descriptors are stored, this step implies more computational resources, slowing down the selection process. We found that regenerating the index when its size is 4 times larger than the original is a reasonable value for this parameter.
- The remaining parameters are shared with FEATMap and, therefore, they will not be discussed again in this section. The reader is referred to Sect. 5.5.1 for further details about their configuration.

6.6.2   Loop Closure Detection

In this case, contrary to Chap. 5 where an indoor sequence was used, we process five outdoor sequences for validating the loop closure detection capabilities of BINMap, namely City Center, New College, KITTI 00, KITTI 05 and KITTI 06, because of the better scalability of BINMap with regard to FEATMap. These sequences are very dynamic and represent several challenging scenarios including important changes in the environment, such as traffic or pedestrians. The reader is referred to Sect. 4.2 for further details about these sequences.

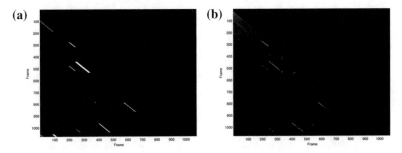

Fig. 6.3 **a** Ground truth loop closure matrix for the New College sequence. **b** Likelihood matrix computed using BINMap

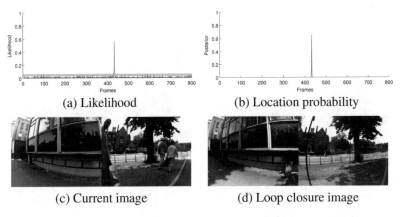

(a) Likelihood   (b) Location probability

(c) Current image   (d) Loop closure image

Fig. 6.4 Example of loop closure detection in the City Center sequence. **a** Likelihood given the current image. **b** Full posterior after the normalization step. Image 971 **c** closes a loop with image 430 **d**. Red and green lines show respectively s_μ and $s_\mu + 2s_\sigma$

Figure 6.3 illustrates the performance of OBIndex for computing the observation likelihood in the New College sequence. The left image is the ground truth while the right image shows the likelihood values obtained for every pair of frames using BINMap. As can be seen, the likelihood is very similar to the ground truth, obtaining high values when real loop closures exist. As in FEATMap, the likelihood is noisier at the beginning of the sequence, which decreases as more images are processed.

An example of loop closure detection in the City Center sequence can be found in Fig. 6.4. In this figure, (a) shows the likelihood computed given the image 971, which is consistent with the posterior shown in (b). Both plots present a high peak around the image 430. Figure 6.4 (c) and Fig. 6.4 (d) are respectively the current robot view and the retrieved location. As can be seen, our approach is able to detect the loop closure situation despite there are changes in the scene.

The performance of the algorithm is evaluated in terms of precision-recall. As in the previous chapter, BINMap is compared against FAB-MAP 2.0 [22]. The binaries of this algorithm are executed using the outdoor visual dictionary and the default

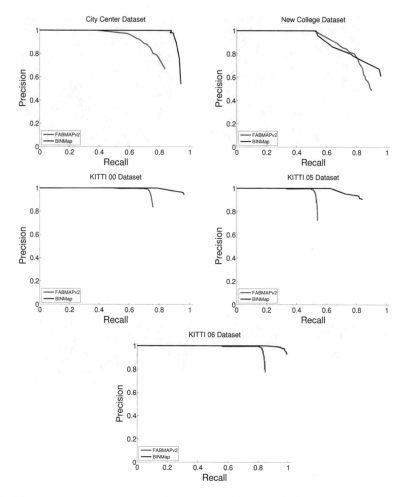

Fig. 6.5 Precision-recall curves for each sequence using BINMap and FAB-MAP 2.0

parameters provided by the authors, and the resulting files of each sequence are processed as explained in Sect. 4.3.1. The resulting precision-recall curves are shown in Fig. 6.5 using BINMap and FAB-MAP 2.0. The BINMap curves result after modifying the threshold for loop acceptance (τ_{loop}). The best results for a 100% of precision are shown in Table 6.2 for an easier understanding of the algorithm performance.

As can be seen, BINMap, as well as FEATMap, outperforms FAB-MAP in all sequences. This is particularly evident for the City Center sequence, where the maximum recall obtained by BINMap is 88.24, in front of the 38.50 reported by FAB-MAP (2.2 times better). This increment of performance remains more or less constant for the KITTI sequences: KITTI 00 (1.5 times better), KITTI 05 (1.9 times better) and KITTI 06 (1.5 times better). For the New College, the maximum recall value is

Table 6.2 Results for the five sequences using BINMap and FAB-MAP 2.0. Precision (Pr) and Recall (Re) columns are expressed as percentages

Sequence	BINMap						FABMAPv2	
	TP	TN	FP	FN	Pr	Re	Pr	Re
City centre	497	676	0	64	**100**	**88.24**	100	38.50
New college	220	656	0	193	**100**	**53.15**	100	51.91
KITTI 00	622	3751	0	168	**100**	**78.73**	100	49.21
KITTI 05	301	2280	0	180	**100**	**62.58**	100	32.15
KITTI 06	228	832	0	41	**100**	**84.76**	100	55.34

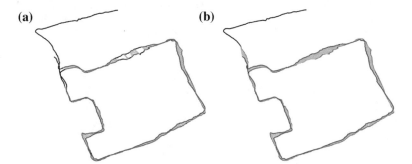

Fig. 6.6 Topological map generated for the City Center sequence. The positions of the images are plotted as black dots. Wherever an image closes a loop with another image, both are labelled with a red dot and linked with a green line. **a** shows the result of BINMap, while **b** shows the ideal map that should be obtained if all the loops present in the sequence were correctly detected

similar to the one obtained using FAB-MAP. We found this sequence very challenging since it presents changes between some semi-indoor and outdoor areas, making more complex the loop closure detection task. As in FEATMap, it was possible to avoid false positives in all cases, while a high number of true positives (TP) and true negatives (TN) were found.

6.6.3 Topological Mapping and Localization

The topological maps obtained for each sequence are presented from Fig. 6.6 to Fig. 6.10. Given that the image positions of these sequences are available and BIN-Map processes each image of the sequence, we can spatially plot the images to produce an easy-to-understand graphical representation of each topological map. Then, when a loop closure is detected, images representing the loop are labelled in red and linked with a green line. Left figures show the map obtained with our approach, while the right figures show the ideal maps that should be obtained if all the loops of the sequences were correctly detected. As can be observed, most loops

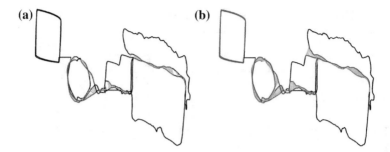

Fig. 6.7 Topological map generated for the New College sequence. The positions of the images are plotted as black dots. Wherever an image closes a loop with another image, both are labelled with a red dot and linked with a green line. **a** shows the result of BINMap, while **b** shows the ideal map that should be obtained if all the loops present in the sequence were correctly detected

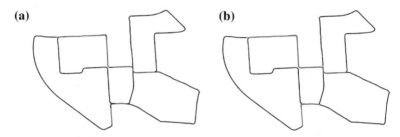

Fig. 6.8 Topological map generated for the KITTI 00 sequence. The positions of the images are plotted as black dots. Wherever an image closes a loop with another image, both are labelled with a red dot and linked with a green line. **a** shows the result of BINMap, while **b** shows the ideal map that should be obtained if all the loops present in the sequence were correctly detected

are detected and the maps are very similar. The resulting maps represent topologically the real scenario, and no false links were added. As commented in the previous section, the map of the New College sequence (Fig. 6.7) is the most different with regard to its ground truth map. This is particularly notorious in the square-shaped part of the map, where no loop closures were found. As a general rule, most part of the true negatives (TN) are produced in the intersections between paths, as shown in Figs. 6.8 and 6.9, or when the vehicle passes by a previously visited place but at a certain distance to the original route, as it is clearly visible in Fig. 6.10. The recall values are lower than the ones obtained in FEATMap, but it is important to note that the sequences used to validate BINMap are longer than the ones used in Chap. 5, augmenting the probability of false detections.

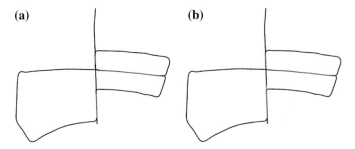

Fig. 6.9 Topological map generated for the KITTI 05 sequence. The positions of the images are plotted as black dots. Wherever an image closes a loop with another image, both are labelled with a red dot and linked with a green line. **a** shows the result of BINMap, while **b** shows the ideal map that should be obtained if all the loops present in the sequence were correctly detected

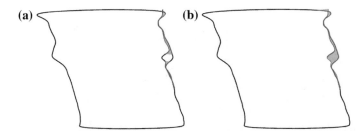

Fig. 6.10 Topological map generated for the KITTI 06 sequence. The positions of the images are plotted as black dots. Wherever an image closes a loop with another image, both are labelled with a red dot and linked with a green line. **a** shows the result of BINMap, while **b** shows the ideal map that should be obtained if all the loops present in the sequence were correctly detected

6.6.4 Computational Times

In this section, we evaluate the performance of BINMap in terms of computational times. We execute the algorithm against the KITTI 00 sequence, which is the longest one considered in this book. To verify the improvement in performance that OBIndex presents against FEATMap for searching descriptors, Fig. 6.11 shows the times needed for searching 650 features per image according to the number of visual words stored in the dictionary. Note that this is critical to reduce the likelihood computation time, which was one of the main drawbacks of FEATMap. As can be seen, when the index contains approximately 1.5 m of visual words, the total search time is only about 50 ms, which can be considered fast in comparison with FEATMap. The searching time initially grows fast, but for larger amounts of features, the growth is far more contained, as shown in the figure.

The computational times invested in each part of the algorithm are shown in Fig. 6.12 and summarized in Table 6.3, where *FAST* refers to the time needed to compute 650 corners, *BRIEF* is the time needed to describe these corners, and

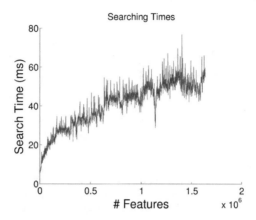

Fig. 6.11 Time required for searching 650 descriptors regarding the number of features stored in the index

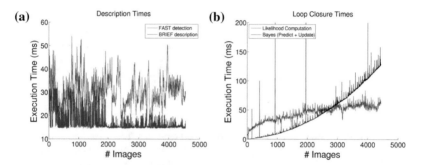

Fig. 6.12 Computational times of BINMap on the KITTI 00 sequence. **a** Computational times for describing an image. **b** Computational times for loop closure detection

Table 6.3 Computational times for BINMap on the KITTI 00 sequence. All times are expressed in milliseconds. The % column indicates the percentage of total time that approximately represents each step

		Mean	%	Std	Max	Min
Image description	FAST	30.10	21.75	5.87	44.17	7.82
	BRIEF	17.18	12.41	4.03	34.68	14.33
Loop closure	Likelihood	45.05	32.55	12.94	83.63	5.75
	Bayes predict	40.02	28.91	35.57	148.18	0.004
	Bayes update	6.07	4.38	13.61	782.24	0.19
	Total	138.42		40.84	1092.9	28.10

the *Likelihood Computation*, *Bayes Predict* and *Bayes Update* rows have the same meaning as in the previous chapter. As can be observed, average FAST detection (30.10 ms) and BRIEF description (17.18 ms) times are substantially shorter than SIFT (1543.30 ms) or SURF approaches (283.60 ms), which represents an important increment of speed of BINMap against FEATMap. Another advantage of BINMap compared to FEATMap is its scalability: OBIndex enables to process more images maintaining a reasonable response time when searching for loop closure candidates, improving the likelihood computation (45.05 ms on average), in contrast to FEATMap (1439.02 ms on average). Note that, according to Fig. 6.12 (b), the Bayes filter steps take even more time than the likelihood computation after processing 3000 images, approximately. BINMap can process an image in 138.42 ms on average according to our experiments, which clearly outperforms FEATMap, which takes, at best, 1759.41 ms.

6.7 Discussion

In this chapter we have introduced BINMap, a topological mapping framework based on a probabilistic loop closure detection algorithm and a discrete Bayes filter, which uses an incremental Bag-of-Binary-Words approach to search for loop closure candidates. Unlike other approaches that make use of visual dictionaries generated offline, in this work we propose an online visual dictionary based on a hierarchical decomposition of the search space by means of a tree. This index can be used with binary descriptors, which improves the speed of the description and searching processes. In our image database approach, named OBIndex, we extend an existing binary indexing algorithm to be used as an online visual dictionary, which, in combination with an inverted index, enables us to obtain loop closure candidates in an efficient way. An extensive set of results have been reported in order to validate BINMap. BINMap has been also compared with FAB-MAP 2.0, outperforming this solution for all sequences.

BINMap has been introduced as a solution to the main drawbacks presented by FEATMap, namely faster image description times and a better scalability. Despite the obvious increment in performance, BINMap can also be improved in terms of precision-recall. The reason is that BINMap uses a single instance of OBIndex for all the images, which means that all the map is searched when a query image is received. As more visual words are inserted into the index, the dispersion among the image candidates augments, increasing the number of false detections. Under these circumstances, in the following chapter, we introduce a hierarchical approach which, instead of searching the whole map, it only considers the locations with an appearance similar to the query image. After that, a more accurate search is performed by means of maintaining an instance of OBIndex for each location. In order to select similar places, a global description method will be shown useful for summarizing the visual content of the images inside a location.

References

1. Cummins, M., Newman, P.: FAB-MAP: probabilistic localization and mapping in the space of appearance. Int. J. Robot. Res. **27**(6), 647–665 (2008)
2. Angeli, A., Filliat, D., Doncieux, S., Meyer, J.A.: A fast and incremental method for loop-closure detection using bags of visual words. IEEE Trans. Robot. **24**(5), 1027–1037 (2008)
3. Fraundorfer, F., Engels, C., Nister, D.: Topological mapping, localization and navigation using image collections. In: IEEE/RSJ International Conference on Intelligent Robots and Systems, pp. 3872–3877 (2007)
4. Galvez-Lopez, D., Tardos, J.: Real-Time loop detection with bags of binary words. In: IEEE/RSJ International Conference on Intelligent Robots and Systems, pp. 51–58 (2011)
5. Sivic, J., Zisserman, A.: Video google: a text retrieval approach to object matching in videos. In: IEEE International Conference on Computer Vision, pp. 1470–1477 (2003)
6. Nister, D., Stewenius, H.: Scalable recognition with a vocabulary tree. IEEE Conference on Computer Vision and Pattern Recognition, vol. 2, pp. 2161–2168 (2006)
7. Lowe, D.G.: Distinctive image features from scale-invariant keypoints. Int. J. Comput. Vis. **60**(2), 91–110 (2004)
8. Bay, H., Tuytelaars, T., Van Gool, L.: SURF: speeded up robust features. In: European Conference on Computer Vision, Lecture Notes in Computer Science, vol. 3951, pp. 404–417 (2006)
9. Calonder, M., Lepetit, V., Strecha, C., Fua, P.: BRIEF : binary robust independent elementary features. In: European Conference on Computer Vision, Lecture Notes in Computer Science, vol. 6314, pp. 778–792 (2010)
10. Leutenegger, S., Chli, M., Siegwart, R.: Brisk: binary robust invariant scalable keypoints. In: IEEE International Conference on Computer Vision, pp. 2548–2555 (2011)
11. Rublee, E., Rabaud, V., Konolige, K., Bradski, G.: ORB: an efficient alternative to SIFT or SURF. IEEE International Conference on Computer Vision, vol. 95, pp. 2564–2571 (2011)
12. Alahi, A., Ortiz, R., Vandergheynst, P.: FREAK : fast retina keypoint. In: IEEE Conference on Computer Vision Pattern Recognition, pp. 510–517 (2012)
13. Yang, X., Cheng, K.T.: Local difference binary for ultrafast and distinctive feature description. IEEE Trans. Pattern Anal. Mach. Intell. **36**(1), 188–94 (2014)
14. Galvez-Lopez, D., Tardos, J.: Bags of binary words for fast place recognition in image sequences. IEEE Trans. Robot. **28**(5), 1188–1197 (2012)
15. Khan, S., Wollherr, D.: IBuILD: incremental bag of binary words for appearance-based loop closure detection. In: IEEE International Conference on Robotics and Automation, pp. 5441–5447 (2015)
16. Garcia-Fidalgo, E., Ortiz, A.: On the use of binary feature descriptors for loop closure detection. In: IEEE Emerging Technologies and Factory Automation, pp. 1–8 (2014)
17. Gionis, A., Indyk, P., Motwani, R.: Similarity search in high dimensions via hashing. In: International Conference on Very Large Data Bases, pp. 518–529 (1999)
18. Salakhutdinov, R., Hinton, G.: Semantic hashing. Int. J. Approx. Reason. **50**(7), 969–978 (2009)
19. Muja, M., Lowe, D.G.: Fast matching of binary features. In: Conference on Computer and Robot Vision, pp. 404–410 (2012)
20. Sparck Jones, K.: A statistical interpretation of term specificity and its application in retrieval. J. Doc. **28**, 11–21 (1972)
21. Rosten, E., Drummond, T.: Machine learning for high-speed corner detection. In: European Conference on Computer Vision, pp. 430–443 (2006)
22. Cummins, M., Newman, P.: Appearance-only SLAM at large scale with FAB-MAP 2.0. Int. J. Robot. Res. **30**(9), 1100–1123 (2011)

Chapter 7
Hierarchical Loop Closure Detection for Topological Mapping

Abstract This chapter describes a novel appearance-based approach for topological mapping called *HTMap* (Hierarchical Topological Mapping), which is based on a hierarchical decomposition of the environment. Images with similar appearances are grouped together in *locations*, taking as a representative of the group the average of the PHOG global descriptors of the represented images, as well as the set of their local features, which are indexed by means of OBIndex (which handles them as explained in the previous chapter). As a main innovation, the algorithm proposes a two-level approach to detect loop candidates: first, the global descriptor of the current image is used to determine the most similar location of the map; next, local image features are employed to determine the most likely image within that location.

7.1 Overview

Most of the recent topological mapping approaches generate dense topological maps, where each input image is introduced as a new node in the map. In these cases, despite the use of different indexing techniques such as the Bag-Of-Words (BoW) schemes [1], the time needed to detect loop closures increases with the number of images. A hierarchical representation of the environment [2], where images which present a similar appearance are grouped together in nodes, can help in these cases, reducing the search space when looking for similar places. To select places, a global descriptor can be useful for summarizing the appearance of a place.

In this chapter, we propose a novel appearance-based approach for topological mapping based on a hierarchical decomposition of the environment called *HTMap* (Hierarchical Topological Mapping). In HTMap, images with similar visual properties are grouped together in *locations*, which are represented by means of an average global descriptor and an instance of OBIndex as image database, working as explained in Sect. 6.2.2. Each image is represented by means of a global descriptor and a set of local features, and this information is used in a two-level loop closure

© Springer International Publishing AG 2018
E. Garcia-Fidalgo and A. Ortiz, *Methods for Appearance-based Loop Closure Detection*, Springer Tracts in Advanced Robotics 122,
https://doi.org/10.1007/978-3-319-75993-7_7

approach, where, first, global descriptors are employed to obtain the most likely nodes of the map and, then, binary image features are used to retrieve the most likely images inside these nodes. This hierarchical scheme enables us to reduce the search space when recognizing places keeping high the representativeness of the map.

As a main contribution of this chapter, we introduce a robust hierarchical loop closure algorithm, which operates in a two-level approach. First, Pyramid Histogram of Oriented Gradients (PHOG) [3] global descriptors are calculated and employed for selecting the locations most similar to the current image, avoiding the need of searching in the whole map and speeding up the retrieval process. Next, binary local features extracted from the current image are used to query the indices of the locations obtained at the previous step in order to find similar images inside the retrieved locations. The scores obtained at the two levels are combined as a likelihood inside a Bayes filter to determine the image most similar to the current image. Two additional contributions are, on the one hand, a scalable method to construct topological maps which employs, as a key component, the above-mentioned hierarchical loop closure algorithm, and, on the other hand, to the best of our knowledge, the use for the first time of the PHOG descriptor for mapping and localization tasks. PHOG represents local image shape and its spatial layout, and was originally devised for image classification. Finally, we performed an extensive evaluation of our approach and a comparison with some state-of-the-art techniques, achieving better results in several public datasets. The accuracy and the sparsity of the generated maps are also discussed.

The chapter is organized as follows: Sect. 7.2 introduces the image description techniques used in our approach, Sect. 7.3 explains the structure of the map generated by HTMap, Sect. 7.4 describes HTMap in detail, Sect. 7.5 reports on the results of the different experiments performed, and Sect. 7.6 concludes the chapter.

7.2 Image Description

In HTMap, images are described using global and local descriptors. When looking for similar places, global descriptors allow us to obtain, in a fast way, a subset of the nodes of the map whose stored images are similar to the current one. Then, local feature descriptors are used to select the most similar images inside the retrieved nodes. An image at time stamp t is described as $I_t = \{G_t, F_t\}$, where G_t is the global descriptor and F_t is the set of local features found in the image. The techniques used for computing G_t and F_t are detailed in this section.

7.2.1 Global Feature Description

As a global representation, we use the Pyramid of Histograms of Orientation Gradients (PHOG) global descriptor [3], which was originally developed for image classification. Despite the fact that it can result less effective for accurate place recognition, this simple descriptor can help when summarizing image information inside a node, as will be shown later. The reader is referred to Sect. 2.2.1.1 and to the original paper [3] for further information about PHOG. In our implementation, we achieve the best results with 60 bins (K) and 3 levels (L), which generates a descriptor of 1260 components. Formally, the global descriptor is defined as $G_t = \{g_0^t, g_1^t, \ldots, g_{1259}^t\}$. According to the original paper, the χ^2 distance exhibits a superior performance when comparing two of these descriptors. Hence, given two PHOG descriptors, G_i and G_j, their distance $d_g(G_i, G_j)$ is defined as:

$$d_g(G_i, G_j) = \sum_{k=0}^{1259} \frac{(g_k^i - g_k^j)^2}{g_k^i + g_k^j}. \tag{7.1}$$

7.2.2 Local Feature Description

Since HTMap uses OBIndex for indexing features at the image level, we compute for each image a collection of FAST features [4] and describe each by an LDB binary descriptor [5]. Besides, we require local features to cover the full image in a more or less uniform way and, to this end, a 4×4 regular grid is defined over the image. The set of LDB descriptors of the n features found at image I_t is defined as $F_t = \{f_0^t, f_1^t, \ldots, f_{n-1}^t\}$. Each of these descriptors is a binary string computed using simple intensity and gradient difference tests on pairwise grid cells within a patch at different spatial granularities. The similarity between two binary strings is computed by means of the Hamming distance. Then, as in (6.6), the distance $d_f(f_p^i, f_q^j)$ between two binary descriptors p and q of, respectively, sets F_i and F_j, is defined as:

$$d_f(f_p^i, f_q^j) = \text{bitsum}(f_p^i \oplus f_q^j).$$

7.3 Map Representation

Our map representation is based on the observation that the appearance between images taken at the same physical place should remain more or less similar. These images can then be grouped together in what we call *locations*. Hence, a location is a group of images of the environment that present some visual similarity. In order to manage the relationships between these locations, the environment is modelled by means of an undirected graph, whose nodes represent the locations in the map and edges represent connectivities between them. Formally, given $I = \{I_0, I_1, \ldots, I_t\}$ as the input sequence of images up to time t, we define our topological map at t as:

$$M_t = (\gamma, \omega), \tag{7.2}$$

where γ is a graph which encodes the topological relationships between locations and ω is the set of existing locations:

$$\omega = \{\ell_0, \ell_1, \ldots, \ell_{c-1}\}, \tag{7.3}$$

where ℓ_i represents the location i and c is the total number of locations. Particularly, the ith location is defined as the tuple:

$$\ell_i = (\zeta_i, \phi_i, \beta_i), \tag{7.4}$$

where $\zeta_i = \{\iota_0, \iota_1, \ldots, \iota_{m-1}\}$ are the indices of the m images associated to the location, ϕ_i is the representative of the location and β_i is an instance of OBIndex built from the images belonging to the location, which is used to retrieve images according to the local binary features found in the current image. When a new image I_t is added to the location, t is added to ζ_i and ϕ_i and β_i are updated accordingly.

Given a query image, the representative descriptor ϕ_i is used to rapidly obtain a measure of similarity between that image and the location i, which is a key step in our hierarchical loop closure algorithm. This representative ϕ_i is computed as the average PHOG descriptor of the images inside the location. Note that this makes sense since images within a location are supposed to present a similar appearance and, therefore, PHOG histograms should be similar as well. Hence, it can be defined as $\phi_i = \{r_0^i, r_1^i, \ldots, r_{1259}^i\}$, where each component is computed as:

$$r_j^i = \frac{\sum_{k=0}^{m-1} g_j^{\iota_k}}{m}. \tag{7.5}$$

After retrieving the most similar locations, we employ local binary descriptors to obtain similar images in these locations. In order to avoid image-to-image comparisons, we make use of the index of local features β_i, which is an instance of OBIndex as explained in Sect. 6.2. Unlike BINMap, HTMap maintains an index of images for each location, avoiding the search within the whole map for each query image.

Figure 7.1 illustrates an example of a map generated using HTMap. Note that our hierarchical decomposition favours long-term tasks. On the one hand, as mentioned previously, it speeds up the loop closure detection process by preventing a search for images throughout the whole map. On the other hand, in order to save storage space in memory, locations can be serialized to disk and loaded on demand if the location is selected as a candidate. Only the representative of the location ϕ_i needs to be maintained in memory for it to be available for the first step of the loop closure detection algorithm.

7.4 Topological Mapping Framework

7.4.1 Algorithm Overview

HTMap builds a visual representation of the environment using a monocular camera, and localizes the robot within this map. Therefore, at time stamp $t - 1$, there exists an *active* location $\ell_a \in \omega$, which can be defined as the current topological position of the robot within the map according to the images up to time $t - 1$. Given the next image I_t, our mapping algorithm tries to determine if there exists a similar location in the map or, otherwise, this image corresponds to an unexplored area of the environment.

Figure 7.2 and Algorithm 7.1 illustrate our topological mapping approach. An initial location ℓ_0, labelled as active, is created including only the first input image I_0. For each new image I_t, global descriptor G_t and local descriptors F_t are computed as explained in Sect. 7.2. These descriptors are then used in the loop closure step to determine if this image comes from an already known place. If positive and the

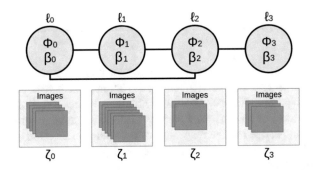

Fig. 7.1 Example of a hierarchical map generated by our approach. The map comprises four locations ℓ_0, ℓ_1, ℓ_2 and ℓ_3 and a loop between ℓ_0 and ℓ_2. Besides the corresponding set of images ζ_i, each location i contains a representative ϕ_i and an index of local features β_i

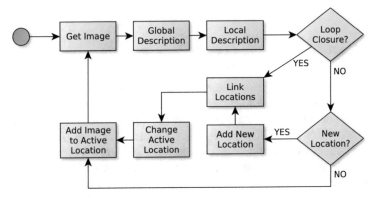

Fig. 7.2 Overview of HTMap

Algorithm 7.1 HTMap: Topological Mapping Framework

1: **procedure** TOPOLOGICAL_MAPPING
2:     **while** there are images **do**
3:         I_t = get_image()
4:         G_t = global_description(I_t)   ▷ PHOG extraction
5:         F_t = local_description(I_t)   ▷ FAST and LDB computation
6:         **if** loop_closure(G_t, F_t, M_t) **then**
7:             ℓ_c = get_loop_closure_location()
8:             link(ℓ_a, ℓ_c, M_t)
9:             $\ell_a = \ell_c$ ▷ Updates the active location
10:        **else**
11:            **if** is_new_location(G_t, ℓ_a) **then**
12:                ℓ_n = create_new_location(M_t)
13:                link(ℓ_a, ℓ_n, M_t)
14:                $\ell_a = \ell_n$ ▷ Updates the active location
15:            **else**
16:                do_nothing()   ▷ The image belongs to ℓ_a
17:            **end if**
18:        **end if**
19:        add_image_to_location(I_t, ℓ_a)
20:    **end while**
21: **end procedure**

---

retrieved location is different to the current active location ℓ_a, the locations are linked in the graph γ in order to register a topological relationship between these places and ℓ_a is updated to point to the loop location. Otherwise, a decision about whether this image belongs to ℓ_a or else is a new place is made. If it can be considered as a new place, a location is added to the map, linked to the current location ℓ_a and labelled as active. In the last step, I_t is associated to ℓ_a, and ϕ_a and β_a are updated by means of, respectively, G_t and F_t.

In the following sections we discuss the hierarchical loop closure detection algorithm and the policy used for determining whether an image belongs to the current active location or is a new location.

7.4.2 Hierarchical Loop Closure Detection

As in the previous solutions, the hierarchical loop closure detection module is based on a discrete Bayes filter, which estimates the probability that the current image closes a loop with a previously seen image associated to an existing location of the map. The approach is outlined in Fig. 7.3 and Algorithm 7.2. In order to avoid false loop closure detections with immediately previous images, the latter are not directly included as loop closure hypotheses as soon as they arrive. Instead, a buffer is used again to store the most recent p images, delaying their publication as loop closure candidates. Consequently, the first step is to release the candidates that could be considered

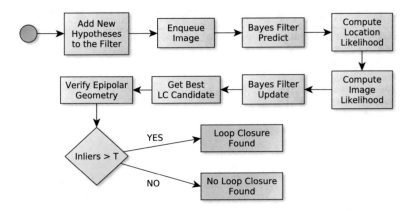

Fig. 7.3 Hierarchical loop closure detection algorithm. LC stands for loop closure

as possible loop closures at the current time step. After that, the current image is enqueued in the buffer and, then, the computation of the likelihood and the Bayes filter update steps are performed. An epipolarity analysis between the current image I_t and the image with the highest probability I_c is performed in order to validate if they can come from the same scene after a camera rotation and/or translation. Matchings that do not fulfil the epipolar constraint are discarded by means of RANSAC. If the number of inliers is above a threshold τ_{ep}, the loop closure hypothesis is accepted; otherwise, it is definitely rejected.

The Bayes filter described below is based on the one described in Chap. 5 after being adapted to be used within a hierarchical approach. Let the pair $L_i^t = \{\ell_j^t, I_i^t\}$ denote the event that image I_t closes a loop with image I_i, which is associated to location ℓ_j at time stamp t, where $i < t$. We also denote $O_t = \{G_t, F_t\}$ as the observation at time t, which comprises the global and local descriptions computed for the current image I_t. Using these definitions, we want to find the previous image I_c whose index satisfies:

$$c = \arg\max_{i=0,\dots,t-p}\{P\left(L_i^t|O_{0:t}\right)\}, \tag{7.6}$$

where $P\left(L_i^t|O_{0:t}\right)$ is the full posterior probability at time t given all previous observations up to time t. As mentioned previously, the most recent p images are not included as hypotheses in the computation of the posterior. Despite the change of notation here due to the hierarchical model of HTMap, the derivation of the filter is the same as the one presented in Sect. 5.4.2 and the reader is referred to that section for further information. Consequently, the full posterior can be written as:

$$P\left(L_i^t|O_{0:t}\right) = \eta P\left(O_t|L_i^t\right) \sum_{j=0}^{t-p} P\left(L_i^t|L_j^{t-1}\right) P\left(L_j^{t-1}|O_{0:t-1}\right), \tag{7.7}$$

Algorithm 7.2 HTMap: Hierarchical Loop Closure Detection

1: **procedure** LOOP_CLOSURE(G_t, F_t, M_t)
2:     add_hypotheses(t)   ▷ Add valid hypotheses at time t
3:     enqueue_image(G_t, F_t)
4:     bayes_filter_predict()
5:     likelihood = compute_likelihood(G_t, F_t, M_t)
6:     bayes_filter_update(likelihood)
7:     c = get_best_candidate()
8:     inliers = epipolar_geometry(F_t, F_c)
9:     **if** inliers > τ_{ep} **then**
10:         **return** true   ▷ Loop closure found
11:     **else**
12:         **return** false   ▷ No loop closure found
13:     **end if**
14: **end procedure**

---

where η is a normalizing factor, $P\left(O_t|L_i^t\right)$ is the observation likelihood, $P\left(L_i^t|L_j^{t-1}\right)$ is the transition model and $P\left(L_j^{t-1}|O_{0:t-1}\right)$ is the posterior distribution computed at the previous time step. The observation model $P\left(O_t|L_i^t\right)$ is computed in two consecutive steps according to the observation pair O_t and using conditional probability properties:

$$P\left(O_t|L_i^t\right) = P\left(G_t, F_t|L_i^t\right) = P\left(G_t|L_i^t\right) P\left(F_t|L_i^t, G_t\right) , \qquad (7.8)$$

where $P\left(G_t|L_i^t\right)$ is determined through the similarity between G_t and the existing locations in the map, and $P\left(F_t|L_i^t, G_t\right)$ is computed searching for similar images inside the retrieved locations.

7.4.2.1   Transition Model

As in the previous solutions, in order to predict the posterior, an evolution model is applied to the probability of loop closure at time $t-1$ using a discretized Gaussian-like function centred at each image j. In HTMap, 90% of the total probability is distributed among eight of the neighbours of j and the remaining 10% is shared uniformly across the rest of loop closure hypotheses according to:

$$\frac{0.1}{\max\{0, t-p-9\}+1} . \qquad (7.9)$$

Note that, unlike FEATMap and BINMap, the size of the Gaussian is eight instead of four. This increases the sensitivity of the filter since the likelihood computation is limited to the images belonging to the locations which are similar enough to the current image I_t.

7.4.2.2 Observation Model

The current observation O_t is included in the filter once the prediction step has been performed. To this end, we make use of our hierarchical representation of the environment, which allows us to calculate the likelihood without the need of computing the similarity between I_t and all the previous images. This likelihood is calculated at two levels: first, global descriptors are used to obtain the locations most similar to the current image, what produces a similarity score for every location in the map. Then, local feature descriptors are searched only in the feature indices at the locations whose score is above a threshold, in order to obtain a similarity score respect to the images stored in those locations.

The goal of the first step is to obtain places in the map with a similar appearance to the current image I_t. To this end, the distance $d_g(\phi_i, G_t)$ between the global descriptor of the image and the representative global descriptor of each location is computed. Next, these distances are converted into a similarity score g_i by means of Eq. 7.10:

$$g_i = 1 - \frac{d_g(\phi_i, G_t) - d_g^{min}}{d_g^{max} - d_g^{min}}, \tag{7.10}$$

where d_g^{min} and d_g^{max} are, respectively, the minimum and the maximum distances resulting for I_t. In order to select the most likely places given the current image, a set of locations ω', whose score is higher than a predefined threshold τ_{llc}, is defined as:

$$\omega' = \{\ell_i \in \omega \mid g_i > \tau_{llc}\}, \tag{7.11}$$

where ω is defined in Eq. 7.2. The final set of candidate locations ω^c is obtained combining ω' with the current active location in the map:

$$\omega^c = \omega' \cup \ell_a. \tag{7.12}$$

The intuition behind the inclusion of the currently active location is that, if a loop was detected at the previous time with an image associated to ℓ_a, it is possible that the current image also closes a loop with an image in ℓ_a, due to the visual similarity between consecutive images. Hence, we update the likelihood of the images associated to ℓ_a irrespective of the score g_i resulting for the images belonging to ℓ_a.

In a second step, additional image similarities are computed in order to find the most likely images in the selected locations. To this end, local binary features of the current image are searched in the feature indices of the retrieved locations ω^c, and a similarity score l_j is computed for every image in each candidate location:

$$l_j = \text{sim\_score}(F_t, \beta_i), \ \forall I_j \in \ell_i, \forall \ell_i \in \omega^c, \tag{7.13}$$

which is a score based on the TF-IDF weighting factor as explained in Sect. 6.2.2 (see Eq. 6.5). Next, the combined similarity score s_j^i of the image j which is stored

---

Algorithm 7.3 HTMap: Likelihood Computation

1: **procedure** COMPUTE_LIKELIHOOD(G_t, F_t, M_t)
2:     $d_g = []$
3:     **for** each location i in ω **do**
4:         $d_g[i] =$ compute_global_dist(ϕ_i, G_t)
5:     **end for**
6:     $d_{max} =$ get_max(d_g)
7:     $d_{min} =$ get_min(d_g)
8:     $g = []$ ▷ Global scores
9:     $\omega^c = [\ell_a]$ ▷ Similar locations to the current image
10:     **for** each location i in ω **do**
11:         $g[i] = 1 - \frac{d_g[i] - d_{min}}{d_{max} - d_{min}}$
12:         **if** $g[i] > \tau_{llc}$ **then**
13:             $\omega^c = \omega^c \cup \ell_i$
14:         **end if**
15:     **end for**
16:     $s = []$ ▷ Combined Scores
17:     **for** each location i in ω^c **do**
18:         **for** each image j in ℓ_i **do**
19:             $l_j =$ sim_score(F_t, β_i)
20:             $s[j] = g[i] \cdot l_j$
21:         **end for**
22:     **end for**
23:     **return** s
24: **end procedure**

---

at location i is defined as the product of global score g_i and the local score l_j:

$$s^i_j = g_i \cdot l_j . \tag{7.14}$$

The likelihood is then calculated, similarly to the previous solutions, according to the following rule:

$$P\left(O_t|L^t_i\right) = \begin{cases} \dfrac{s^i_j - s_\sigma}{s_\mu} & \text{if } s^i_j \geq s_\mu + s_\sigma \\ 1 & \text{otherwise} \end{cases} , \tag{7.15}$$

being, respectively, s_μ and s_σ the mean and the standard deviation of the set of scores. Notice that, by means of Eq. 7.15, given the current observation O_t, only the most likely images update their posterior, and, in this case, the function is more selective than in FEATMap or BINMap. The likelihood computation is formally stated in Algorithm 7.3. To finish, the full posterior is normalized in order to obtain a probability density function after incorporating the observation into the filter.

7.4.3 New Location Policy

Once the current image has been determined not to close a loop with any existing location, the dissimilarity between ℓ_a and I_t is evaluated in order to determine if the image has been taken from the same physical place represented by the current location. To this end, the distance between the representative of the node and the global descriptor of the image is computed and contrasted against a threshold τ_{nn}, so that if $d_g(\phi_a, G_t) < \tau_{nn}$, then I_t is associated to ℓ_a. This parameter plays a key role with regard to the sparsity of the map: the higher the value of τ_{nn}, the lower the number of nodes, but more images are associated to each location. In Sect. 7.5, we will evaluate how the quality and accuracy of the generated maps varies according to this parameter.

7.5 Experimental Results

In this section we evaluate HTMap from different points of view. The section is organized as follows: first, we discuss about the configuration of the parameters; secondly, the performance of our hierarchical loop closure detection algorithm is evaluated; next, the ability of HTMap to create topological maps from the environment is assessed; then, the quality of the maps with regard to their sparsity is analysed; finally, HTMap computational times are discussed.

7.5.1 Parameter Configuration

In the same way as for FEATMap and BINMap, in this section we discuss the different parameters that way affect the performance of HTMap. The algorithm was configured using the parameters indicated in Table 7.1. Most part of the parameters of HTMap are also present in the previous solutions and have been already discussed. Therefore, they are not reviewed again here and the reader is referred to Sects. 5.5.1 and 6.6.1 for further information about their configuration. Two additional parameters are considered in HTMap:

- The new location threshold (τ_{nn}), which directly affects the sparsity of the resulting maps: the higher the value, the lower the number of resulting locations. This is a critical parameter for HTMap execution and will be deeply discussed in Sect. 7.5.4.
- The loop closure node threshold (τ_{llc}), which affects the number of selected nodes when searching for loop closure candidates. The higher this value, the higher the number of nodes selected, but it implies to search in more nodes at image loop closure level. We set this value to 0.65, which is enough to maximize the recall rates avoiding to search in all nodes.

Table 7.1 Parameters for HTMap execution

Parameter	Value
Branching factor (K)	16
Maximum leaf size (L)	150
Number of search trees (T)	4
Rebuild threshold (R)	4
Keypoints per image (n)	650
Nearest neighbour ratio (ρ)	0.8
Previous images discarded (p)	30
Number of inliers (τ_{ep})	Varying
New location threshold (τ_{nn})	0.15
Loop closure node threshold (τ_{llc})	0.65

Note that, unlike previous solutions where the parameter τ_{loop} governed loop acceptance decisions, in HTMap there are no thresholds for accepting final loop closures. Therefore, τ_{ep} becomes a critical parameter and has been used for plotting the precision-recall curves shown in the next section.

7.5.2   Loop Closure Detection

For the loop closure evaluation, we use the same sequences as in the previous chapter for the evaluation of BINMap: City Centre, New College, KITTI 00, KITTI 05 and KITTI 06. As usual, the reader is referred to Sect. 4.2 for further details about these sequences.

Figure 7.4 shows the observation likelihood computed for the New College sequence using HTMap. As illustrated in the figure, the likelihood results to be less noisy than for BINMap (see Fig. 6.3), what reduces the number of false positives, as will be shown later. This effect is inherent to the likelihood computation scheme used in HTMap, since it selects several locations of the map by means of the similarity between global descriptors, and then only the images inside these locations are taken into account during the likelihood computation. Conversely, the loops are shown as peaks lower than in BINMap, but they are enough for increasing the performance of the solution.

We evaluate the performance of the hierarchical loop closure approach for recognizing previously seen places. The assessment is performed in terms of precision-recall. As usual, in our validation tests we are particularly interested in the maximum

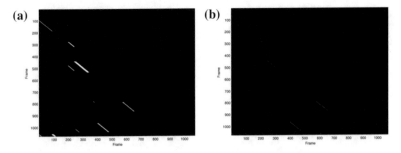

Fig. 7.4 **a** Ground truth loop closure matrix for the New College sequence. **b** Likelihood matrix computed using HTMap

recall that can be achieved at 100% precision, what implies no false positives in any case.

Since HTMap is based on a hierarchical combination of both BoW and global schemes, we want to verify its performance in comparison with two solutions of each of these paradigms executed alone. Therefore, we have performed a comparison between FAB-MAP 2.0 [6] and SeqSLAM [7], which can be considered as state-of-the-art approaches of, respectively, each paradigm. FAB-MAP 2.0 is evaluated using the binaries provided by the authors with default parameters and the results are processed as explained in Sect. 4.3.1. Conversely, SeqSLAM is validated according to the procedure explained in Sect. 4.3.2.

The precision-recall curves for each sequence are shown in Fig. 7.5. In each plot, the curves were plotted modifying τ_{ep} in HTMap, p in FAB-MAP 2.0 and the loop closure acceptance threshold in SeqSLAM. For an easier understanding of the curves, best results for a 100% of precision are also shown in Table 7.2. As can be seen in the figure, the area under the curve (AUC) of HTMap is higher than the corresponding curves for the other solutions, outperforming them in all sequences. According to our experiments, SeqSLAM is usually able to obtain higher recall at 100% of precision than FAB-MAP 2.0, except for New College, where results are very similar. The performance of our approach is specially high in KITTI 00 and KITTI 06 sequences, where more than 90% of recall can be obtained at 100% of precision. The maximum recalls obtained for the other sequences are 79.68% for City Centre, 73.60% for New College and 75.88% for KITTI 05, which are very high in comparison with the other solutions. HTMap achieves a higher recall than BINMap in all sequences, except for the City Centre sequence (see Table 6.2). In general, the hierarchical loop closure approach used by HTMap enhances the performance of BINMap in terms of precision-recall.

Following the same notation as the one used in Sect. 6.6.3 for plotting the generated maps of BINMap, the detected loops in each corresponding sequence are shown from Figs. 7.6, 7.7, 7.8, 7.9 and 7.10. Left figures show the map obtained with our approach, while the right figures show the ideal maps that should be obtained if all the loops of the sequences were correctly detected. As in BINMap, note that most part of

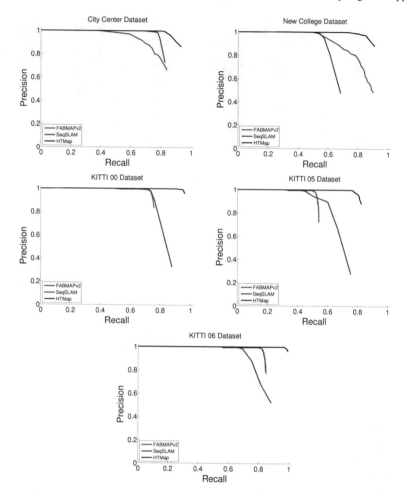

Fig. 7.5 Precision-recall curves for each sequence using HTMap, FAB-MAP 2.0 and SeqSLAM

Table 7.2 Results for the five sequences using HTMap, FAB-MAP 2.0 and SeqSLAM. Precision (Pr) and Recall (Re) columns are expressed as percentages

Sequence	HTMap						FABMAPv2		SeqSLAM	
	TP	TN	FP	FN	Pr	Re	Pr	Re	Pr	Re
City centre	447	676	0	114	**100**	**79.68**	100	38.50	100	68.98
New college	304	660	0	109	**100**	**73.60**	100	51.91	100	49.39
KITTI 00	712	3752	0	77	**100**	**90.24**	100	49.21	100	67.04
KITTI 05	365	2280	0	116	**100**	**75.88**	100	32.15	100	41.37
KITTI 06	261	832	0	832	**100**	**97.03**	100	55.34	100	64.68

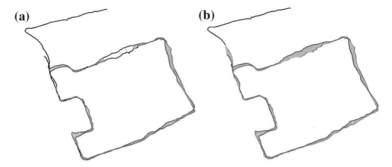

Fig. 7.6 Appearance-based loop closure results for the City Centre sequence. The positions of the images are plotted as black dots. Wherever an image closes a loop with another image, both are labelled with a red dot and linked with a green line. **a** shows the result of HTMap, while **b** shows the ideal map that should be obtained if all the loops present in the sequence were correctly detected

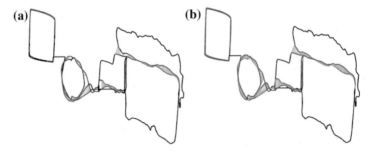

Fig. 7.7 Appearance-based loop closure results for the New College sequence. The positions of the images are plotted as black dots. Wherever an image closes a loop with another image, both are labelled with a red dot and linked with a green line. **a** shows the result of HTMap, while **b** shows the ideal map that should be obtained if all the loops present in the sequence were correctly detected

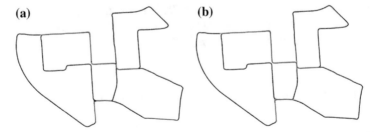

Fig. 7.8 Appearance-based loop closure results for the KITTI 00 sequence. The positions of the images are plotted as black dots. Wherever an image closes a loop with another image, both are labelled with a red dot and linked with a green line. **a** shows the result of HTMap, while **b** shows the ideal map that should be obtained if all the loops present in the sequence were correctly detected

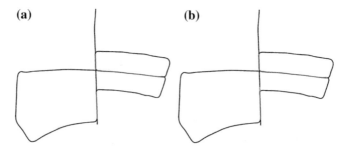

Fig. 7.9 Appearance-based loop closure results for the KITTI 05 sequence. The positions of the images are plotted as black dots. Wherever an image closes a loop with another image, both are labelled with a red dot and linked with a green line. **a** shows the result of HTMap, while **b** shows the ideal map that should be obtained if all the loops present in the sequence were correctly detected

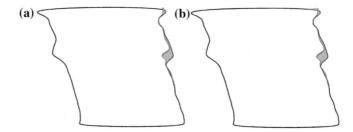

Fig. 7.10 Appearance-based loop closure results for the KITTI 06 sequence. The positions of the images are plotted as black dots. Wherever an image closes a loop with another image, both are labelled with a red dot and linked with a green line. **a** shows the result of HTMap, while **b** shows the ideal map that should be obtained if all the loops present in the sequence were correctly detected

the existing loops are detected, specially in the KITTI 00 and KITTI 05 sequences. HTMap is able to succeed in cases where BINMap is not, e.g. when the vehicle passes by a previously visited place but at a certain distance of the original route. This is specially evident in Figs. 7.7 and 7.10.

7.5.3   Topological Mapping and Localization

The same sequences used in the previous experiments were also employed to validate our framework regarding mapping tasks. To this end, the topological maps obtained at 100% of precision are shown from Figs. 7.11, 7.12, 7.13, 7.14, and 7.15. For each sequence, a random colour has been assigned to each of the locations of the map. All images associated to this location during the generation of the map are then labelled using the corresponding colour. We want to verify that the images are tagged with the same colour when revisiting an already known place. As can be seen, this is true in most cases, and it is specially evident in the City Centre sequence, where a large

Fig. 7.11 Generated
topological map at 100% of
precision for the City Centre
sequence. Images belonging
to the same location are
labelled with the same
colour. The total number of
generated locations is 18

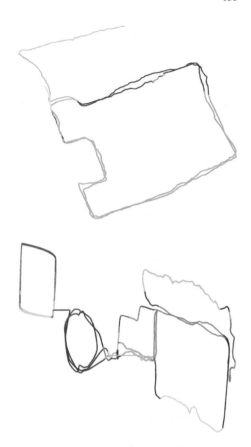

Fig. 7.12 Generated
topological map at 100% of
precision for the New
College sequence. Images
belonging to the same
location are labelled with the
same colour. The total
number of generated
locations is 86

loop is found. In this case, despite the map contains a few locations, HTMap does not get confused and images corresponding to already visited places are assigned to the corresponding node. The total number of locations obtained for each sequence is 18 for City Centre, 86 for New College, 32 for KITTI 00, 19 for KITTI 05 and 20 for KITTI 06.

The effectiveness of the solution in terms of precision-recall with regard to the number of locations of the map varies depending on the sequence. We verify the accuracy of the maps with regard to the number of locations in the following section.

7.5.4   Sparsity

The number of locations of a map with regard to the number of images is defined as its sparsity [2]. The lower the number of locations, the higher the sparsity, since more images are assigned to the same place. However, more sparsity does not imply more

Fig. 7.13 Generated
topological map at 100% of
precision for the KITTI 00
sequence. Images belonging
to the same location are
labelled with the same
colour. The total number of
generated locations is 32

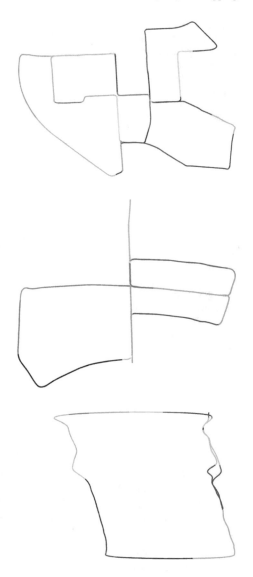

Fig. 7.14 Generated
topological map at 100% of
precision for the KITTI 05
sequence. Images belonging
to the same location are
labelled with the same
colour. The total number of
generated locations is 19

Fig. 7.15 Generated
topological map at 100% of
precision for the KITTI 06
sequence. Images belonging
to the same location are
labelled with the same
colour. The total number of
generated locations is 20

accurate maps, since the number of locations affects the loop closure detection. In
the first level of our loop closure approach, more location candidates are taken into
account, what can induce false loop detections. Moreover, a larger sparsity makes the
binary index of each location contain more visual words, sharing the scores among
more loop closure candidate images. A good mapping technique must balance the
number of locations and the ability of detecting previously seen places. In this section,
we want to evaluate the effect of the sparsity in the performance of HTMap.

Fig. 7.16 Recall of HTMap
according to the number of
locations at 100% of
precision and τ_{ep} set to 150

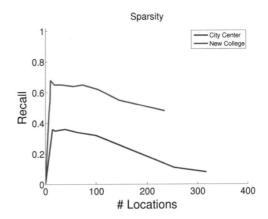

The parameter with a major influence on the sparsity of the maps is τ_{nn}: the higher the value of τ_{nn}, the lower the number of locations (see Sect. 7.4.3). Since we are interested in avoiding false positives, we fix the parameter τ_{ep} to 150, which is enough to ensure 100% of precision in all sequences. Then, we execute our approach varying τ_{nn} and the number of locations and recalls are observed. This experiment has been performed over the City Centre and the New College sequences, since they have approximately the same length and have been taken at the same frame rate. This is to ensure that the number of generated locations is independent of the length of the sequence. Note that using a low value of τ_{ep} can lead to a higher recall, but in this case we are only interested in the relation existing between the number of locations and the recall produced by HTMap under these conditions.

The results are shown in Fig. 7.16. As can be seen, a high number of locations does not imply better performance: from approximately 100 locations, the recall starts to decrease for both sequences. A number of locations between 10 and 80 are enough to guarantee the best recall values, being the recall more or less stable in this interval (≈ 0.34 for City Centre and ≈ 0.65 for New College). This is an interesting point, taking into account that having more locations in the map could imply higher computational load during the first step of our loop closure approach, where the most likely locations are retrieved. Note that minimum sparsity is equivalent to perform loop closure detection at exclusively the image level. This proves that our grouping approach helps during the detection of previously seen places and validates the ability of the PHOG global descriptor to summarize the visual information that characterizes a place.

7.5.5 Computational Times

In this section, we evaluate the performance of HTMap in terms of computational time. To this end, we execute our approach over the KITTI 00 sequence using the

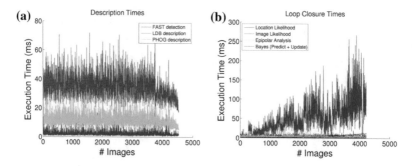

Fig. 7.17 Computational times of HTMap on the KITTI 00 sequence. **a** Computational times for describing an image. **b** Computational times for loop closure detection

parameters that gave us the maximum recall in the previous experiments, generating a total number of 32 locations from a set of 4541 images. We have measured the execution time of the different parts of our algorithm in this sequence, and the results are shown in Fig. 7.17 and summarized in Table 7.3, where *PHOG* is the time needed to perform a global description of the image, *FAST* includes the keypoint detection and the selection of the best n corners, and *LDB* is the time needed to compute the binary description of these keypoints. Regarding the loop closure detection times, *Location Likelihood* refers to the time required to perform the first step of our hierarchical loop closure algorithm, *Image Likelihood* involves the computation of the second step of our approach, *Bayes Predict* is the time needed to make a prediction in the filter and *Bayes Update* is the time needed to update the filter using the computed likelihood. Finally, *Epipolar Analysis* includes matching features and the computation of the fundamental matrix.

As can be observed, times are substantially shorter than including the likelihood computation. Regarding the image description, the computation of PHOG, which took 11.4 ms on average, is even faster than finding features using FAST and describing them using LDB, which took together 37.4 ms. The total time needed for describing an image is 48.8 ms on average, which is much faster than describing an image using SIFT or SURF, as could be expected. The highest execution time corresponds to the image likelihood computation. This time increases with the number of nodes, since more locations must be searched. This effect can be reduced increasing τ_{llc} for selecting less location candidates, which will reduce the recall, although it could be enough depending on the environment. Nonetheless, the maximum time measured is 267.3 ms, which can be considered as a reasonable result. The computation of the location likelihood is very fast despite the increment of the nodes in the map, taking 3.1 ms on average. In short, HTMap can process an image in 109.3 ms on average according to our experiments, which outperforms, for instance, the state-of-the-art algorithm FAB-MAP 2.0, where, according to the experiments presented in [6], only the SURF feature extraction takes 400 ms, approximately.

The main improvement of performance of HTMap with regard to BINMap has to do with the recall values, since the computational times are quite similar, as can

Table 7.3 Computational times for HTMap on the KITTI 00 sequence. All times are expressed in milliseconds. The % column indicates the percentage of total time that approximately represents each step

		Mean	%	Std	Max	Min
Image description	PHOG	11.40	10.43	3.50	41.60	4.90
	FAST	34.90	31.93	7.80	72.00	16.70
	LDB	2.50	2.10	2.10	34.90	0.470
Loop closure	Location likelihood	3.10	2.84	3.40	16.90	0.10
	Image likelihood	48.90	44.74	37.00	267.30	0.01
	Bayes predict	3.90	3.57	3.50	19.40	0.01
	Bayes update	0.40	0.37	0.20	2.10	0.01
	Epipolar analysis	4.20	3.84	1.50	23.20	1.50
	Total	109.30		38.37	477.40	23.70

be seen in the results. Regarding the image description times, the LDB descriptor is faster to compute than the BRIEF descriptor used in BINMap. However, in HTMap, a PHOG descriptor is computed, making the complete description time of an image very similar in both approaches. The likelihood computation time in HTMap is a bit higher than in BINMap, due to the two steps needed by the hierarchical scheme. Conversely, the Bayes steps take less time in HTMap. The epipolar analysis time has been included in the HTMap results since this step is always executed in this approach, unlike the other solutions, where this step is only performed when an image attains a probability higher than a threshold.

7.6 Discussion

In this chapter we have introduced a new topological mapping approach, HTMap, based on a hierarchical place recognition technique. Instead of generating a dense map, where all the images correspond to a node in the final topology, our approach builds a hierarchical decomposition of the environment, where images with similar properties are grouped together to form locations. Each location is represented by means of an average PHOG global descriptor and an instance of OBIndex, which is based on a BoW scheme that can be built online. This map representation is very useful for reducing the search space when searching for loop closures. Then, as a key component of our topological mapping technique, we have also presented a hierarchical loop closure detection method. First, given the current image, PHOG global descriptors are used to obtain the most likely places in the map. After that, local binary features are used to obtain the most likely images belonging to the places retrieved previously. These scores are combined as a likelihood in a discrete Bayes filter. We have verified the utility of the PHOG descriptor in place recognition tasks, showing that it can be very helpful for summarizing the visual information that a place presents. We have validated our approach using several public datasets, and compared our results with SeqSLAM and FAB-MAP 2.0, verifying that our combination of global and BoW approaches performs better than two solutions of each paradigm alone.

According to the experimental results, the hierarchical scheme used in HTMap allows us to improve the recall values obtained with BINMap, since the algorithm selects from the map locations similar to the query image for detecting loop closures, instead of searching in the whole map. HTMap is even faster than BINMap according to our results and can be easily used in long-term tasks, what makes HTMap an interesting option for topological mapping.

References

1. Sivic, J., Zisserman, A.: Video google: a text retrieval approach to object matching in videos. In: IEEE International Conference on Computer Vision, pp. 1470–1477 (2003)
2. Korrapati, H., Mezouar, Y.: Vision-based sparse topological mapping. Robot. Auton. Syst. **62**(9), 1259–1270 (2014)
3. Bosch, A., Zisserman, A., Munoz, X.: Representing shape with a spatial pyramid kernel. In: ACM International Conference on Image and Video Retrieval, pp. 401–408 (2007)
4. Rosten, E., Drummond, T.: Machine learning for high-speed corner detection. In: European Conference on Computer Vision, pp. 430–443 (2006)
5. Yang, X., Cheng, K.T.: Local difference binary for ultrafast and distinctive feature description. IEEE Trans. Pattern Anal. Mach. Intell. **36**(1), 188–94 (2014)
6. Cummins, M., Newman, P.: Appearance-only SLAM at large scale with FAB-MAP 2.0. Int. J. Robot. Res. **30**(9), 1100–1123 (2011)
7. Milford, M., Wyeth, G.: SeqSLAM: visual route-based navigation for sunny summer days and stormy winter nights. In: IEEE International Conference on Robotics and Automation, pp. 1643–1649 (2012)

Chapter 8
Fast Image Mosaicking Using Incremental Bags of Binary Words

Abstract This chapter introduces a fast and multi-threaded algorithm for image mosaicking called *BIMOS* (Binary descriptor-based Image MOSaicking) as another example of task where appearance-based loop closure detection is of utmost importance, as it is for vision-based topological mapping. Actually, an image mosaicking process can be seen as a particular case of topological mapping given that the alignment of the images considered, which can be seen as the topology of the image sequence, has to be determined to generate the image composite. To this end, BIMOS makes use of OBIndex to find overlapping pairs. BIMOS has been validated using image sequences from several kinds of environments.

8.1 Overview

This book has introduced several topological mapping algorithms which are based on visual loop closure detection techniques. However, topological mapping is not the only research area in which it is necessary an efficient place recognition method. A related problem is image mosaicking, where the topological relationships between the images need to be determined in order to correctly align them. These relationships are usually represented by means of a graph, what leads to see image mosaicking as an special case of topological mapping. The topology estimation of the environment is usually a time-demanding process, specially when the number of images is high. This is mainly due to the lack of efficient structures for indexing images and the use of slow image description algorithms such as SIFT [1] or SURF [2]. In this regard, the OBIndex indexing scheme introduced in Chap. 6 can be useful for detecting overlapping pairs.

In this chapter, we propose a novel image approach, named BIMOS (Binary descriptor-based Image MOSaicking),[1] which can produce seamless mosaics on different scenarios and camera configurations in a reasonable amount of time. More precisely, we introduce a multi-threaded architecture for image that allows us to decouple the strategic steps involved in the process, speeding up the time required

---

[1] http://github.com/emiliofidalgo/bimos.

© Springer International Publishing AG 2018
E. Garcia-Fidalgo and A. Ortiz, *Methods for Appearance-based Loop Closure Detection*, Springer Tracts in Advanced Robotics 122,
https://doi.org/10.1007/978-3-319-75993-7_8

to estimate the final topology. To find overlapping candidates, BIMOS employs OBIndex, as explained in Chap. 6, which is based on a Bag-Of-Words (BoW) scheme that is built in an online manner. Our approach takes advantage of the use of the ORB detector and descriptor [3] to accelerate the image description process. BIMOS is validated under different environments and camera configurations, showing that it can be used on several scenarios producing coherent results.

The chapter is organized as follows: Sect. 8.2 introduces the concept of image and the latest related works, Sect. 8.3 presents the motion model used in BIMOS, Sect. 8.4 describes the scheme of BIMOS, Sect. 8.5 reports the experimental results obtained, and Sect. 8.6 concludes the chapter.

8.2 Background

In the last decades, cameras have been widely used for collecting information from the environment. When a robot is equipped with a camera, it is usually of interest to obtain a large visual representation of the operating area, which can be used for close-up inspection, for localization and even for navigation tasks. Since the field-of-view of conventional cameras is limited, image techniques have been developed for building a larger view of the surveyed area. Mosaicking is then defined as the process of stitching images together to provide a wide-area image of the scene. Generally, an image algorithm consist of two main steps:

- *image alignment*, also known as image registration, which is the process of over-laying two or more images of the same scene taken at different times and/or from different viewpoints into a common reference frame [4, 5], usually referred to as the *mosaic frame*, and,
- *image blending*, which is in charge of rendering the images together in a final seamless mosaic.

In this chapter, we are particularly interested in the image alignment step leaving image blending techniques out of the scope of this book. In BIMOS, we use a generic algorithm for blending the resulting mosaics. The reader is referred to [6] for an extensive review of image blending techniques.

When a stream of images is the only source of information available, the alignment can be performed concatenating the estimated transformations between consecutive pairs of images. This simple method is called *pairwise alignment*. The problem is that this technique rapidly tends to accumulate registration errors, providing a wrong trajectory of the camera, which produces misalignments in the final mosaic. Finding correspondences between non-consecutive images is a key step to correct the trajectory of the vehicle and to reduce this drift. Despite the trajectory estimated by means of pairwise alignment can be used to infer overlapping pairs, it is better to employ another source of information not affected by the accumulated error. In our case, we use an appearance-based loop closure detection technique to detect if an image has been previously seen. After detecting overlaps between non-consecutive images, *global alignment* methods can be used to optimize the trajectory of the vehicle. This iterative process of matching and optimization is called *topology estimation* and usually lasts until no more overlapping pairs can be found [7]. For obvious reasons, graph theory is very useful for defining and representing topologies [8, 9].

The quality and the time needed to obtain the final topology are directly related to the method used for describing images and the ability for finding overlapping pairs. With regard to image description, most part of image approaches make use of SIFT or SURF, due to their invariance properties to illumination, scale and rotation. Recently there has been a growing interest in the development of binary descriptors, such as BRIEF [10], BRISK [11], ORB [3], FREAK [12] or LDB [13], due to their lower computational cost.

In order to detect the topological relationships between images, a frame-to-frame comparison approach can be used only when the number of images is low. As it grows, this approach becomes unfeasible and an indexing scheme is needed for searching overlapping pairs in an efficient way. A BoW approach [14], such as the one presented in Chap. 6, is of application here.

Image drew the attention of the robotics community some years ago, specially for mapping areas using down-looking cameras, a configuration adopted by most of the systems presented so far. However, it is less usual to find solutions that make use of forward-looking cameras, as described in [15] for inspecting vessels using a Micro-Aerial Vehicle (MAV) and in [16] for inspecting hydroelectric dams using an Autonomous Underwater Vehicle (AUV). Image algorithms have been usually validated for a single environment. As will be seen later, BIMOS has been tested against several kind of scenarios and camera configurations.

Image has been extensively used for underwater environments and different tasks. For instance, Gracias et al. [17] introduced a visual navigation approach for underwater robots based on image. A visual servoing controller is employed for controlling the trajectory of the vehicle, which is specified directly over the mosaic. Pizarro [18] and Madjidi [19] present solutions to the problem of large-area global by underwater vehicles. More recently, Elibol et al. [20] devise a global alignment method based on graph theory, which works on the mosaic frame and does not require non-linear optimization. Ferreira et al. [21] introduce a real-time algorithm based on a SLAM system which relies on the BRIEF descriptor.

Several image techniques involving MAVs have recently emerged have been validated using these kind of vehicles. Kekec et al. [22] present a real-time algorithm for creating mosaics from aerial images. In this approach, the Separating Axis Theorem (SAT) is used for detecting image intersections. Bulow and Birk [23] introduce a fast and robust method for visual odometry which is then used to generate large photo maps from images taken from a MAV. Botterill et al. [24] developed an aerial image algorithm using an offline BoW scheme.

8.3 Motion Estimation

The model employed to estimate the camera motion plays a key role in the image
registration process. BIMOS assumes that either the scene is planar or the distance
from the camera to the scene is high enough so as to neglect depth changes. It is also
assumed that the camera is more or less perpendicular to the scene and that keeps at
a more or less constant distance.

Under these conditions, two overlapping images I_i and I_j are related by a homography, a linear transformation represented by a 3×3 matrix $^i H_j$ such that:

$$p_i = {}^i H_j \, p_j \, , \tag{8.1}$$

where p_i and p_j are two corresponding image points from, respectively, I_i and I_j,
expressed in homogeneous coordinates. Despite BIMOS can deal with any kind of
homography, we approximate the motion of the camera by a simpler model using a
similarity transformation, which has four degrees of freedom, comprising rotation,
translation and scale. $^i H_j$ is expressed as:

$$
{}^i H_j =
\begin{pmatrix}
s \cos\theta & -s \sin\theta & t_x \\
s \sin\theta & s \cos\theta & t_y \\
0 & 0 & 1
\end{pmatrix}
=
\begin{pmatrix}
a & -b & c \\
b & a & d \\
0 & 0 & 1
\end{pmatrix} ,
\tag{8.2}
$$

where s is the scale, θ the rotation angle and (t_x, t_y) the translation vector. The estimation of any of these homographies starts by matching corresponding points between
images. Maximum Likelihood Estimation Sample Consensus (MLESAC) [25] is
next used robustly to minimize the reprojection error for Eq. 8.2 and discard outliers
(see Eq. 8.7 later).

Finally, for the case of a path of images $I_i, I_{k_1}, \ldots, I_{k_m}, I_j$, the associated transformation that relates frames I_i and I_j is computed by concatenating the corresponding
relative homographies:

$$
{}^i H_j = {}^i H_{k_1}{}^{k_1} H_{k_2} \ldots {}^{k_{m-1}} H_{k_m}{}^{k_m} H_j \, .
\tag{8.3}
$$

8.4 Image Mosaicking Using Binary Descriptors

In this section, we describe BIMOS, whose architecture is outlined in Fig. 8.1.
Inspired by some recent SLAM solutions such as PTAM [26] and ORB-SLAM [27],
BIMOS comprises four threads that run in parallel, each one in charge of a strategic
step of the algorithm. This configuration allows us to decouple the execution of the
different parts of BIMOS, reducing the time needed to generate a mosaic. All threads
interact with a shared structure called *mosaic graph*, which is used to manage the

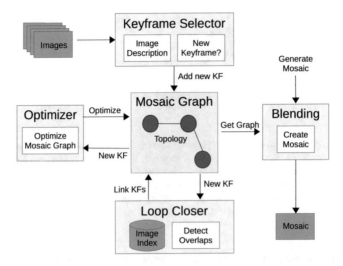

Fig. 8.1 BIMOS architecture. The four threads (in blue) interact with a shared structure called *mosaic graph* (in green). The arrows indicate the main actions performed by the different components

topology of the environment and the synchronization mechanisms between threads. The *keyframe selector* thread, which is the entry to the system, describes the input images and decides if they should be part of the final image composite. The *loop closer* thread detects overlapping image pairs and the *optimizer* thread reduces the global misalignment in the graph performing a bundle adjustment process. Finally, the *blending* thread is responsible for generating the final image mosaic. BIMOS is ready to work online using a Robot Operating System (ROS) topic through which it processes images on demand, contrary to most algorithms, which work offline. In the following sections, we describe the building blocks of BIMOS.

8.4.1   Mosaic Graph

The topology of the environment represents the relationships that exist between the images conforming the mosaic. In our approach, the topology is modelled by means of an undirected graph, where nodes represent the image subset that will lead to the final mosaic and links represent the overlaps among them. In BIMOS, the selected images are called *keyframes*.

The *mosaic graph* is a key component of BIMOS. It manages the graph that represents the topology of the environment and provides mechanisms to ensure the exclusive access of the different threads to this graph. In order to create the final mosaic, keyframes need to be aligned according to a common selected frame, referred to as the *mosaic frame*, as stated before. Then, each keyframe is associated

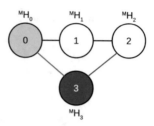

Fig. 8.2 A simple example of a mosaic graph comprising four keyframes and their corresponding absolute homographies. Keyframe 0, coloured in green, is the mosaic frame and, therefore, $^M H_0$ is the identity. Keyframe 3 is the last keyframe inserted in the graph. The node and the link with keyframe 2, in blue, were added by the *keyframe selector* thread. The link with keyframe 0, in red, was added by the *loop closer* thread after detecting an overlap between images. The *optimizer* thread adjusts the absolute homographies of the graph

to an absolute homography $^M H_i$, which relates the correspondent keyframe i with the mosaic frame M. In this work, since BIMOS processes images on demand and the graph is updated as new images arrive, the first keyframe is always selected as the reference frame of the mosaic. Its absolute homography is thus the identity matrix. A relative homography is also associated to each link, which will be used during the pose-graph optimization step.

Several threads of the system modify concurrently the mosaic graph: the *keyframe selector* thread inserts new keyframes in the graph, the *loop closer* thread links keyframes as it detects overlapping image pairs and the *optimizer* thread globally adjusts the absolute homographies $^M H_i$. The mosaic graph structure and its use are illustrated in Fig. 8.2.

8.4.2 Keyframe Selection

This component is responsible for describing the input images and deciding which ones are useful for building the final mosaic. First of all, the ORB [3] algorithm is used to detect and describe a set of keypoints in the image. We use ORB due to its good tolerance to rotations [27], instead of BRIEF [10] and LDB [13], used by BINMap and HTMap. However, note that BIMOS is descriptor-independent and any detector-binary descriptor combination can be used. Besides, to favour accurate estimation of the image transformations, a minimum number of features (3000) is requested to be found, and they are required to cover the full image in a more or less uniform way over a 4×4 regular grid defined over the image.

Instead of using all the input images, we apply a keyframe selection policy in order to discard images which are not deemed to provide a significant contribution to the mosaic, avoiding unnecessary drift during the alignment process. This contribution is measured as the amount of overlap between the current image and the last keyframe inserted in the graph, so that the higher the overlap the less relevant is the image. Note

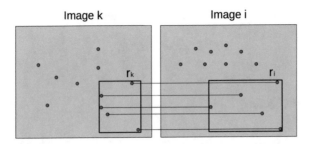

Fig. 8.3 An simple example of bounding rectangle computation. Red points represent the keypoints found in each image (I_k and I_i, respectively). The surviving inliers after computing the homography between the images ($^kH_i^*$) are indicated with a black line. The bounding rectangles at each image are shown in black

that this policy is different to the one employed in Chap. 5 since, in this application, we require to validate the visual overlap between the two images and not only if they represent the same place. More specifically, we compute the homography $^kH_i^*$ between the current image i and the last inserted keyframe k. Given the resulting set of inliers, we obtain the coordinates of the corresponding points in each image. We then calculate the minimal up-right bounding rectangle for each point set, formally r_j for image j, and, next, we evaluate the percentage of overlap that this bounding rectangle represents in the image. A simple example of the computation of these bounding rectangles is illustrated in Fig. 8.3. The overlap is expressed as follows:

$$O_j = \frac{\text{area}(r_j)}{w_j \times h_j},\qquad(8.4)$$

where the function area($\cdot$) computes the area of the bounding rectangle and w_j and h_j are, respectively, the width and the height of image j in pixels. In order to take a final decision, the overlap between the images is computed as:

$$^kO_i = min(O_k, O_i).\qquad(8.5)$$

Then, if there are enough inliers and the overlap ^kO_i is high enough, the current image i is stored as a potential keyframe. Otherwise, the last potential keyframe found is added to the mosaic graph, and the transformation from the current image to the new keyframe is recomputed. This policy allows us to ensure that, despite we are discarding several images, there exists a minimum overlap between consecutive keyframes and the topology is not broken in different parts.

When an image i is added as keyframe into the mosaic graph, it is linked with the previous keyframe. The link is associated to the relative homography $^kH_i^*$. Its absolute homography is initialized concatenating the absolute homography of the previous keyframe with $^kH_i^*$. Then, following the notation used in this chapter, if the image i is added as the keyframe $k+1$ in the graph, $^kH_i^*$ becomes $^kH_{k+1}$ and, consequently, the initial absolute homography is given by:

$$^M H_{k+1} = {}^M H_k{}^k H_{k+1} \, . \tag{8.6}$$

8.4.3 Loop Closing

This thread detects which keyframes close a loop with previously added keyframes. To this end, we use OBIndex as explained in Chap. 6. This component maintains an instance of OBIndex, which indexes all the keyframes defined up to the current time. When a new keyframe is added, it is searched in the index, obtaining a list of candidates sorted from highest to lowest visual similarity. Next, each candidate is evaluated in descending order, computing the homography with the current keyframe. If the number of resulting inliers is higher than a certain threshold, a link to the corresponding keyframes is incorporated into the graph. Otherwise, the process finishes and, if exists, the next keyframe is processed.

Since consecutive images are linked by default, we want to find overlapping pairs at farther distances, which is of prime importance during the optimization step. To achieve this, keyframes are not directly indexed as soon as they are processed. Instead, a buffer is used to store the most recent keyframes, delaying their publication as overlapping candidates for the following keyframes.

8.4.4 Optimization

Despite the efforts for accurately estimating the topology such as robust homography computation, keyframe selection and loop closure detection, alignment errors still arise, resulting into globally inconsistent mosaics. To correct this problem, this component is in charge of performing a bundle adjustment step to jointly minimize the global misalignment induced by the current absolute homographies. The reprojection error function is defined as follows:

$$\varepsilon = \sum_i \sum_j \sum_{k=1}^n \| p_i^k - (^M H_i)^{-1}\, {}^M H_j\, p_j^k \| + R(^M H_j) +$$
$$\| p_j^k - (^M H_j)^{-1}\, {}^M H_i\, p_i^k \| + R(^M H_i) \, , \tag{8.7}$$

where i and j are two images related by a link, n is the total number of resulting inliers when computing the relative homography $^i H_j$, (p_i^k, p_j^k) are the corresponding points for the inlier k, $^M H_i$ and $^M H_j$ are the absolute homographies for, respectively, images i and j, and $R(^M H_i)$ and $R(^M H_j)$ are regularization terms. These terms prioritize homographies with scale closer to 1 during the optimization, since BIMOS assumes that the camera moves at a more or less constant distance from the scene.

These terms are defined as follows:

$$R(^M H_i) = \gamma \left(a^2 + b^2 - 1\right) = \gamma \left((s \cos\theta)^2 + (s \sin\theta)^2 - 1\right) \qquad (8.8)$$

where γ is a regularization factor, s and θ are, respectively, the scale and the orientation contained in the homography, and a and b are defined in Eq. 8.2.

To reduce the influence of outliers, we optimize, instead of Eq. 8.7, a Huber robust error function:

$$h(\varepsilon) = \begin{cases} |\varepsilon|^2 & \text{if } |\varepsilon| \leq 1 \\ 2|\varepsilon| - 1 & \text{if } |\varepsilon| > 1 \end{cases} . \qquad (8.9)$$

The system of non-linear equations is solved by means of the Levenberg–Marquardt algorithm using the Ceres Solver library[2] and the absolute homographies available so far as a starting point. Usually a few iterations are needed to achieve convergence.

In BIMOS, a short optimization is executed periodically after the insertion of a certain number of keyframes in the graph, limiting the optimization to a maximum of 30 s and 50 iterations. This parameter is of prime importance in the performance of the algorithm, since excessive optimizations may slow down the process. Just before the blending step, a longer optimization (for a maximum of 600 s and 1000 iterations) is also performed to finally adjust the absolute homographies. Note that, despite the different convergence criteria, both optimizations adjust the absolute homographies of the whole mosaic.

8.4.5 Blending

This last component makes use of a multi-band blending algorithm [28] to create the final seamless mosaic. This step is an adaptation of the *stitching* module implemented in the OpenCV library, which includes seam finding and exposure compensation. In BIMOS, this component runs as a thread on demand, which permits generating mosaics at different moments along the process.

8.5 Experimental Results

BIMOS has been validated under different operating conditions and using several datasets. The results obtained for each dataset are summarized in Table 8.1, indicating the total number of images in the input set (#Imgs), the number of keyframes selected by BIMOS (KFs), the execution times corresponding to the different phases of the algorithm —global alignment (Alig), global optimization (Opt), blending (Blend)

---

[2]http://ceres-solver.org/.

Table 8.1 Summary of the experimental results. Times are expressed in seconds and errors are expressed in pixels

| | | | BIMOS | | | | | | Old approach [15] | | |
| | | | Ex. times | | | | Rep. error | | Ex. times | Rep. error | |
Seq	#Imgs	KFs	Alig	Opt	Blend	Total	Avg	Std	Total	Avg	Std
Vall1	201	80	14.4	0.4	38.7	53.5	2.2	2.2	187.5	2.7	3.0
Vall2	2504	335	330.1	0.4	2876.2	3206.7	8.1	14.8	9042.9	7.9	10.2
MAV	137	88	3.7	0.2	21.5	25.3	1.8	1.8	108.6	2.2	2.1
Air1	71	32	5.3	0.5	68.7	74.5	4.3	6.6	224.4	4.1	3.2
Air2	840	336	106.3	2.7	1008.4	1118.0	6.3	6.8	4766.7	5.9	5.3

Fig. 8.4 (top) Mosaic resulting for the *Vall1* dataset. (bottom) Topology estimated by BIMOS. Each keyframe is indicated using a red circle, and the mosaic frame is labelled by a green triangle

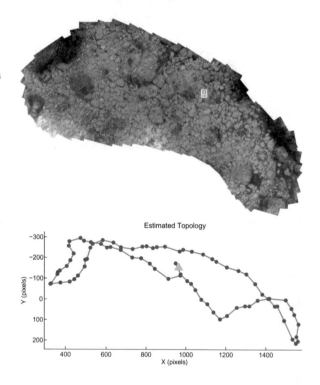

and the total time needed to build the mosaic (Total)— and, finally, the average and standard deviation of the reprojection error calculated using all the correspondences with the resulting set of homographies (Avg, Std). Note that the global alignment time also includes the small optimizations produced during the estimation of the topology. Given that BIMOS is the evolution of a previously released solution [15], we also include in the table the execution times and the reprojection error of this algorithm in order to show the performance improvement that BIMOS represents against this

Fig. 8.5 (top) Mosaic resulting for the *Vall2* dataset. (bottom) Topology estimated by BIMOS. Each keyframe is indicated using a red circle, and the mosaic frame is labelled by a green triangle

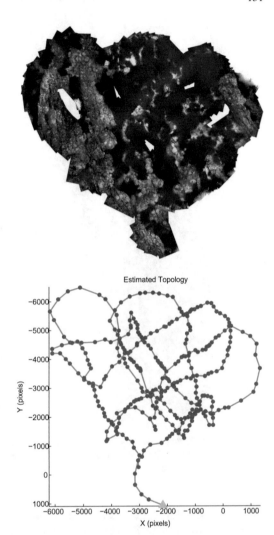

solution. BIMOS goes a step further introducing a more generic approach which is based on a multi-threaded architecture and a new image selection policy, making it faster than this previous solution.

For the case of BIMOS, and differently from the other solutions introduced in this book, all experiments have been performed on a desktop computer fitted with an Intel Core i7 at 4.4Ghz processor and 32GB of RAM memory.

As a first experiment, we use an underwater dataset whose images come from the Valldemossa harbour seabed (Mallorca, Spain) and a hand-held down-looking camera. The dataset consists of 201 images of 320 × 180 pixels, which comprises a large loop, what allows us to validate the ability of our algorithm for recognizing previously seen places. A total number of 80 images were selected by BIMOS, leading

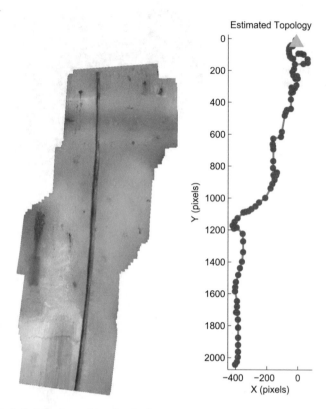

Fig. 8.6 (left) Mosaic resulting for the image sequence collected by a *MAV*. (right) Topology estimated by BIMOS. Each keyframe is indicated using a red circle, and the mosaic frame is labelled by a green triangle

to the final mosaic and the topology shown in Fig. 8.4. Despite the reprojection error is similar to the one obtained by our previous approach, the time needed to complete the mosaic is only 53.57 s in front of 187.52, which implies an execution time 3.5 times shorter.

The second dataset, also recorded at Valldemossa harbour, is a more challenging environment in the sense that it comprises a *Posidonia* meadow, characterized by a self-similar texture and vegetation in continuous motion. A total number of 2504 images were obtained, covering an area of approximately 400 m². BIMOS selects 335 keyframes, producing the mosaic and the topology shown in Fig. 8.5. As in the previous experiment, we obtain a coherent mosaic in less time than our previous approach.

The third dataset was recorded using a MAV designed for vessel visual inspection [29], which was fitted with a 752 × 480-pixel/58°-lens uEye UI-1221LE camera running at 10 Hz. The MAV was flying at a more or less constant distance from the

Fig. 8.7 (top) Mosaic
resulting for the *Air1*
sequence. (bottom) Topology
estimated by BIMOS. Each
keyframe is indicated using a
red circle, and the mosaic
frame is labelled by a green
triangle

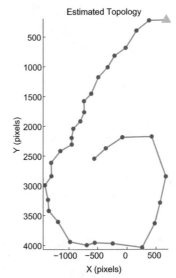

scene (1.5 m). Note that due to the more aggressive dynamics of the MAV and the
front-looking camera configuration, this is a more challenging situation than the
Valldemossa dataset. A total number of 137 images were captured, from where
BIMOS selected 88 as keyframes. As in the previous experiment, the reprojection
error is lower than for our previous approach. However, the most interesting result
has to do with the execution time of BIMOS, which is, according to our results, 4.3
times faster than our previous algorithm. The resulting mosaic and the estimated
topology are shown in Fig. 8.6.

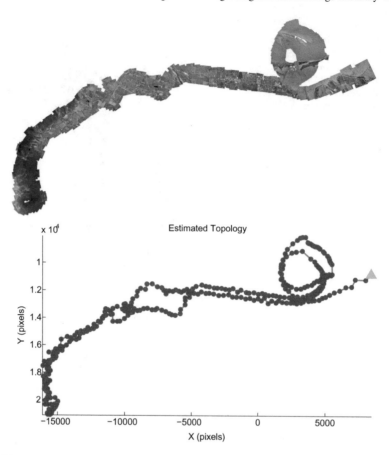

Fig. 8.8 (top) Mosaic resulting for the *Air2* sequence. (bottom) Topology estimated by BIMOS. Each keyframe is indicated using a red circle, and the mosaic frame is labelled by a green triangle

As a fourth experiment, we employ an aerial image sequence taken at high altitude. This dataset, provided by the original authors, was taken using a bottom-looking camera attached to a teleoperated aerial vehicle [24]. This is also a challenging scenario because of the aggressive movement of the vehicle, what makes the camera be far from perpendicular to the scene sometimes. We have considered two sequences from this dataset, comprising, respectively, 71 and 840 images of size 800×533 pixels, corresponding to areas covering several kilometres. Each sequence is identified in Table 8.1 as *Air1* and *Air2*. The corresponding mosaics and the estimated topologies are shown, respectively, in Figs. 8.7 and 8.8. As the other experiments, BIMOS is faster than our previous approach keeping a similar reprojection error.

8.6 Discussion

In this chapter, we have described a novel approach for generating mosaics from images. Our scheme, named BIMOS, is based on a multi-threaded architecture which allows us to decouple the different parts of the algorithm, speeding up the process. The topology of the environment is modelled by means of an undirected graph. To find the overlapping pairs in an efficient way, this graph is created using OBIndex, which is built online. We have validated our approach under different operating conditions, obtaining coherent mosaics in all cases. BIMOS is an example of application where place recognition is also needed to model the environment.

References

1. Lowe, D.G.: Distinctive image features from scale-invariant keypoints. Int. J. Comput. Vis. **60**(2), 91–110 (2004)
2. Bay, H., Tuytelaars, T., Van Gool, L.: SURF: speeded up robust features. In: European Conference on Computer Vision, Lecture Notes in Computer Science, vol. 3951, pp. 404–417 (2006)
3. Rublee, E., Rabaud, V., Konolige, K., Bradski, G.: ORB: an efficient alternative to SIFT or SURF. IEEE Int. Conf. Comput. Vis. **95**, 2564–2571 (2011)
4. Szeliski, R.: Image alignment and stitching: a tutorial. Found. Trends® Comput. Graph. Vis. **2**(1), 1–104 (2006)
5. Zitova, B., Flusser, J.: Image registration methods: a survey. Image Vis. Comput. **21**(11), 977–1000 (2003)
6. Prados, R., Garcia, R., Neumann, L.: Image Blending Techniques and their Application in Underwater Mosaicing. Springer, Berlin (2014)
7. Elibol, A., Gracias, N., Garcia, R.: Efficient Topology Estimation for Large Scale Optical Mapping, vol. 82. Springer, Berlin (2012)
8. Sawhney, H.S., Hsu, S., Kumar, R.: Robust video mosaicing through topology inference and local to global alignment. In: European Conference on Computer Vision, pp. 103–119 (1998)
9. Marzotto, R., Fusiello, A., Murino, V.: High resolution video mosaicing with global alignment. In: IEEE Conference on Computer Vision and Pattern Recognition, vol. 1, pp. I–692–I–698 (2004)
10. Calonder, M., Lepetit, V., Strecha, C., Fua, P.: BRIEF: binary robust independent elementary features. In: European Conference on Computer Vision, Lecture Notes in Computer Science, vol. 6314, pp. 778–792 (2010)
11. Leutenegger, S., Chli, M., Siegwart, R.: BRISK: binary robust invariant scalable keypoints. In: IEEE International Conference on Computer Vision, pp. 2548–2555 (2011)
12. Alahi, A., Ortiz, R., Vandergheynst, P.: FREAK: fast retina keypoint. In: IEEE Conference on Computer Vision and Pattern Recognition, pp. 510–517 (2012)
13. Yang, X., Cheng, K.T.: Local difference binary for ultrafast and distinctive feature description. IEEE Trans. Pattern Anal. Mach. Intell. **36**(1), 188–94 (2014)
14. Sivic, J., Zisserman, A.: Video Google: a text retrieval approach to object matching in videos. In: IEEE International Conference on Computer Vision, pp. 1470–1477 (2003)
15. Garcia-Fidalgo, E., Ortiz, A., Bonnin-Pascual, F., Company, J.P.: A mosaicing approach for vessel visual inspection using a micro aerial vehicle. In: IEEE/RSJ International Conference on Intelligent Robots and Systems (2015)
16. Ridao, P., Carreras, M., Ribas, D., Garcia, R.: Visual inspection of hydroelectric dams using an autonomous underwater vehicle. J. Field Rob. **27**(6), 759–778 (2010)

17. Gracias, N., van der Zwaan, S., Bernardino, A., Santos-Victor, J.: Mosaic-based navigation for autonomous underwater vehicles. J. Ocean. Eng. **28**(4), 609–624 (2003)
18. Pizarro, O., Singh, H.: Toward large-area mosaicing for underwater scientific applications. J. Ocean. Eng. **28**(4), 651–672 (2003)
19. Madjidi, H., Negahdaripour, S.: Global alignment of sensor positions with noisy motion measurements. IEEE Trans. Robot. **21**(6), 1092–1104 (2005)
20. Elibol, A., Garcia, R., Gracias, N.: A new global alignment approach for underwater optical mapping. Ocean Eng. **38**(10), 1207–1219 (2011)
21. Ferreira, F., Veruggio, G., Caccia, M., Zereik, E., Bruzzone, G.: A real-time mosaicking algorithm using binary features for ROVs. In: Mediterranean Conference on Control and Automation, pp. 1267–1273 (2013)
22. Kekec, T., Yildirim, A., Unel, M.: A new approach to real-time mosaicing of aerial images. Rob. Auton. Syst. **62**(12), 1755–1767 (2014)
23. Bulow, H., Birk, A.: Fast and robust photomapping with an unmanned aerial vehicle (UAV). In: IEEE/RSJ International Conference on Intelligent Robots and Systems, pp. 3368–3373 (2009)
24. Botterill, T., Mills, S., Green, R.: Real-time aerial image mosaicing. In: IVCNZ, pp. 1–8 (2010)
25. Torr, P.H., Zisserman, A.: MLESAC: a new robust estimator with application to estimating image geometry. Comput. Vis. Image Und. **78**(1), 138–156 (2000)
26. Klein, G., Murray, D.: Parallel tracking and mapping for small AR workspaces. In: IEEE/ACM International Symposium on Mixed and Augmented Reality (ISMAR), pp. 225–234 (2007)
27. Mur-Artal, R., Tardos, J.D.: ORB-SLAM: tracking and mapping recognizable features. In: Workshop on Multi-View Geometry in Robotics (RSS) (2014)
28. Burt, P.J., Adelson, E.H.: A multiresolution spline with application to image mosaics. ACM Trans. Graph. **2**(4), 217–236 (1983)
29. Bonnin-Pascual, F., Ortiz, A., Garcia-Fidalgo, E., Company, J.P.: A micro-aerial platform for vessel visual inspection based on supervised autonomy. In: IEEE/RSJ International Conference on Intelligent Robots and Systems, pp. 46–52 (2015)

Chapter 9
Conclusions and Future Work

Abstract This chapter summarizes the work done and discusses on the next steps to be undertaken as future work, to further improve the results presented in this book.

9.1 Summary

Given the importance of mapping in autonomous mobile robotics, this book has mainly addressed the problem of developing methods for topological mapping using cameras as a sensor. Topological maps present several benefits in front of the classic metric approaches, specially in tasks without accuracy needs, as discussed in Chap. 2. A wide range of sensors have been used for topological mapping, but all of them (including cameras) are influenced by noise. Therefore, using only raw sensor measurements for mapping tasks leads us to obtain inconsistent representations of the environment. Due to this reason, loop closure detection, understood as the ability of a robot to correctly determine that it has returned to a previously visited place, is a fundamental component in most modern mapping systems to reduce the uncertainty of the maps. The topological approaches introduced in this book heavily rely on loop closures, and, hence, several novel vision-based loop closure detection algorithms have been proposed. Loop closure detection, also known as place recognition, is also an important step in other computer vision areas such as image mosaicking, and, therefore, the approaches presented in this book can be of interest to this end. As an example of application, we have also proposed an image mosaicking algorithm based on one of the appearance-based loop closure detection algorithms introduced in this book.

The performance of a vision-based topological mapping algorithm is mainly influenced by the image description and indexing methods employed. In this regard, Chap. 3 of this book surveys the most recent appearance-based topological mapping approaches emerged the last years. Exploring the work of preceding researchers has allowed us to determine several open research topics which inspired the different contributions of this work.

As a first mapping solution, Chap. 5 has introduced an appearance-based approach for topological mapping and localization named FEATMap (Feature-based

© Springer International Publishing AG 2018

E. Garcia-Fidalgo and A. Ortiz, *Methods for Appearance-based Loop Closure Detection*, Springer Tracts in Advanced Robotics 122,
https://doi.org/10.1007/978-3-319-75993-7_9

Mapping). FEATMap is based on a loop closure detection algorithm which uses local invariant features for image description. To efficiently search for loop closure candidates, matchings between the current image and previously visited images are determined using an index of features based on a set of randomized kd-trees. A novel map refinement strategy has been also presented to remove spurious nodes in the final topology. According to the results obtained, FEATMap exhibits a better performance than FAB-MAP 2.0. However, it invests a significant amount of time in image description and presents scalability issues as more nodes are inserted in the map.

In order to solve the issues of FEATMap, a Bag-of-Words (BoW) scheme, which quantizes the features according to a reference visual dictionary, has been considered of particular relevance. In detail, our interest has been: (1) exploit the benefits of binary descriptors and, (2) avoid the training step usually needed for building these visual dictionaries, and so generate the dictionary online. In Chap. 6, we have introduced OBIndex (Online Binary Image Index), an image index based on an incremental Bag-of-Binary-Words approach and inverted files to obtain similar image candidates. Next, this index has been used as a key component in a probabilistic topological mapping framework called BINMap (Binary Mapping). The experiments performed have shown that BINMap achieves a better performance than FAB-MAP 2.0 and, besides, solves the issues presented by FEATMap.

Despite the good performance exhibited by BINMap, another solution, called HTMap (Hierarchical Topological Mapping) has been proposed in Chap. 7. As a main innovation, HTMap builds a hierarchical representation of the environment: images with similar visual properties are grouped together in locations. Locations are represented by means of an average global descriptor and an instance of OBIndex for indexing the images associated to the node. Then, loop closure detection is performed in two steps: first, global descriptors are used to obtain similar candidate nodes and, next, binary descriptors of the current image are searched in the indices of the selected nodes to finally obtain a similar image candidate. This hierarchical decomposition of the environment improves the scalability of the solution and favours long-term tasks. According to the experimental results, HTMap presents a better performance than FAB-MAP 2.0 and SeqSLAM, and further improves the recognition rates achieved by BINMap.

Table 9.1 summarizes the main features of each of the approaches introduced in this book. In comparison with other state-of-the-art solutions like FAB-MAP 2.0 or

Table 9.1 Summary of the approaches proposed in this book. More '*' means better performance regarding the corresponding attribute

Approach	Descriptor type	Descriptor scope	Indexing method	Scalability	C. times
FEATMap	Real-valued	Local	Rand. kd-trees	*	*
BINMap	Binary	Local	Online BoW	**	**
HTMap	Real-valued/binary	Local/global	Hier. online BoW	***	***

SeqSLAM, all the solutions proposed in this book exhibit a better loop closure performance according to the classical precision-recall metrics obtained. They can be used to generate visual topological maps of the environment. Moreover, none of them requires the typical training phase of BoW approaches, as they have been devised to allow the robotic platform to adapt to its environment. Conversely, BINMap and HTMap represent pioneering solutions using binary features for vision-based topological mapping, which still keeps as a research area to explore, as stated in Chap. 3.

As discussed previously, place recognition is not an exclusive process of topological mapping. There exist other research areas where the ability of recognizing previously seen places is of prime importance, such as, for instance, for image mosaicking. Therefore, in order to demonstrate that the appearance-based techniques developed in this book can be used for other tasks, in Chap. 8 a multi-threaded image mosaicking algorithm has been proposed, which makes use of OBIndex to find overlapping pairs between images. BIMOS has been validated under different environments and camera configurations, producing coherent mosaics in all cases in a reasonable amount of time.

9.2 Future Work

As a future work, the following tasks are planned to be carried out:

- Several improvements can be applied to all the topological mapping approaches presented in this book in order to further improve their performance. Initially, several components are susceptible of being parallelized in a Graphics Processing Unit (GPU) or in multi-processor environments, which will speed up their execution. That is the case, for instance, of the Bayes filters and the image description steps. Next, despite the topological nature of the solutions, adding metrical information to the edges about the spatial relationships between the involved places will be useful for navigation tasks. We are also intent to explore the topological mapping approaches in larger scenarios in order to further validate their scalability and robustness.
- OBIndex is a key component of BINMap, HTMap and BIMOS. Due to this reason, these solutions would benefit from the improvement of the performance of OBIndex. In this regard, adding information about the spatial arrangement and the co-occurrence of the words in a visual dictionary will improve the effectiveness when searching for similar images. Another issue to overcome has to do with limiting the size of the dictionary. Despite the good general performance of OBIndex, a method for purging visual words would be also helpful for long-term tasks.
- Regarding BIMOS, there exist several improvements that can be investigated. First of all, since the main bottleneck of the algorithm is the blending step, exploring other techniques for image blending to speed up BIMOS execution is an interesting task that we plan to tackle. Furthermore, instead of optimizing the whole graph, performing a local optimization when a loop closure is detected will be beneficial for saving computation time.

Printed in the United States
By Bookmasters